LET'S FIND IT!

ALL ABOUT MAPS AND GLOBES
for Young Scientists

LET'S FIND IT!

ALL ABOUT MAPS AND GLOBES
for Young Scientists

HERON
BOOKS

Published by
Heron Books, Inc.
20950 SW Rock Creek Road
Sheridan, OR 97378

heronbooks.com

———————————

Special thanks to all the teachers and students who
provided feedback instrumental to this edition.

———————————

At Heron Books, we think learning should be engaging and fun. It should be hands-on and allow students to move at their own pace.

To facilitate this we have created a learning guide that will help any student progress through this book, chapter by chapter, with confidence and interest.

Get learning guides at
heronbooks.com/learningguides.

For teacher resources,
such as a final exam, email
teacherresources@heronbooks.com.

We would love to hear from you!
Email us at *feedback@heronbooks.com.*

IN THIS BOOK

CHAPTER 1

WHAT ARE MAPS FOR?

A map is a picture of a place as it looks from above. You can use it to find out where things are.

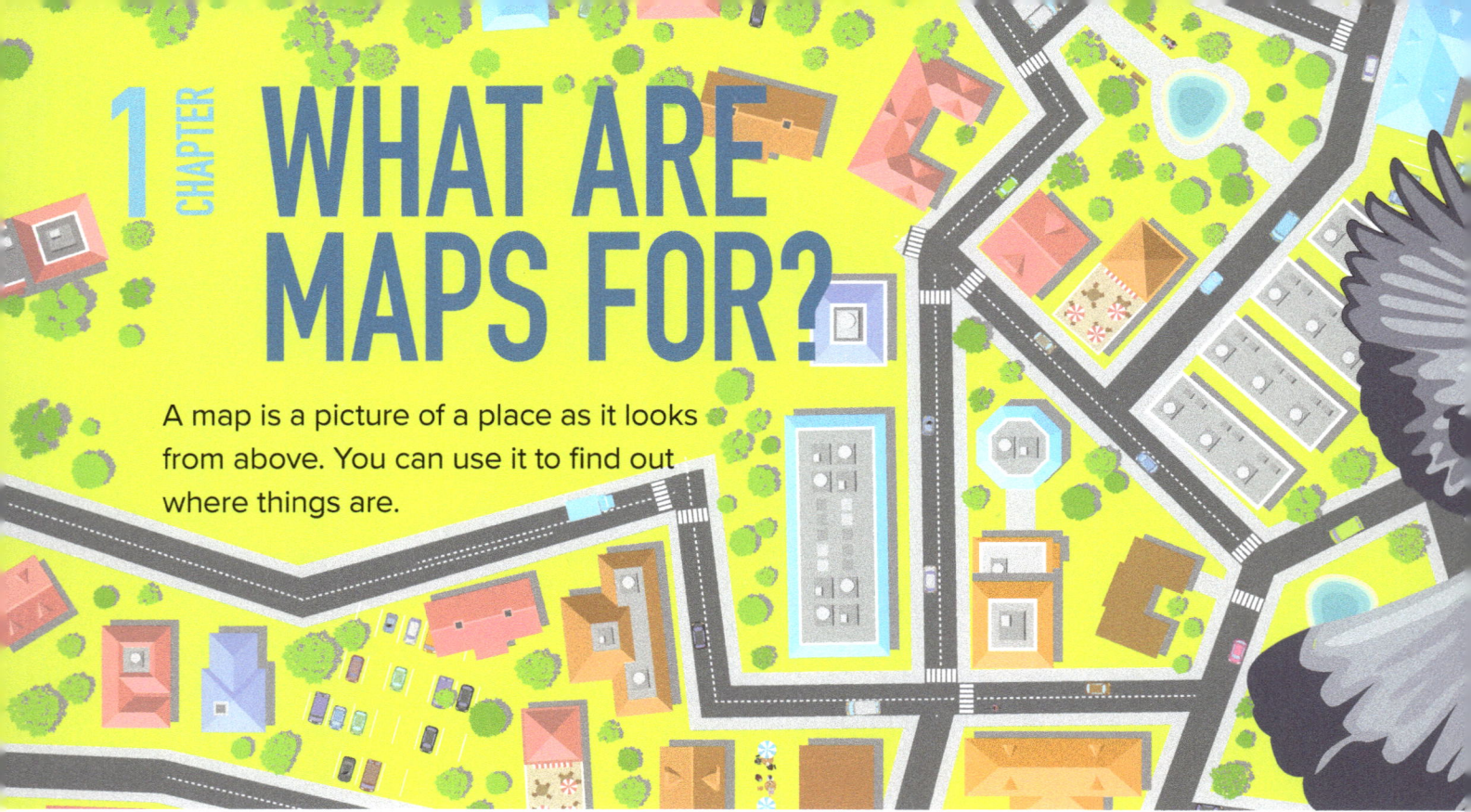

HOW ARE MAPS USED?

When you think about it, people use maps a lot.

MUSEUM MAP

Zoos and museums use maps to help people find the things they want to see.

Shopping malls often have large maps to help people find places in the mall.

Maps are helpful because they give us

PARK MAP

People often use maps on their phones to show them how to get where they want to go.

Amusement parks like Disneyworld use maps to show visitors where different rides are.

When you're hiking in a forest, or biking on a trail, a map can keep you from getting lost.

information in a way that is easy to see.

Sometimes writers draw maps to
make stories easier to follow.

Maps give us information. They tell us about the world and all the places in it. We can use them to find our way, and also answer questions.

How far is New York from here? Do we have time to stop and see it?

Where is Idaho?

Are England and France near one another?

Are there volcanoes in Japan?

What is the world's largest country?

Using a map, you can find new places you might like to learn about or visit someday.

When you have the right map and know how to use it, you can find answers to questions like these and many, many more.

So let's get busy learning how maps work!

2 CHAPTER SCALE

As you probably know, any map is much smaller than the place it is a picture of. If you wanted to make a map of something, how could you go about showing a large place on a fairly small piece of paper?

SHOWING A LARGE PLACE ON A SMALL MAP

Mia's parents planned to get some new furniture for their living room. She decided to help them figure out how to make everything fit by drawing them a map.

They will use this to figure out what furniture to get and where to put it.

Mia started by measuring the size of the room. This information would help her make everything on her map the right size.

4

She found out that the room was 14 feet long.

Then she thought about how long to make the 14-foot wall on her paper, which was only 10 inches in length. She decided to make the wall 7 inches long.

This meant that 7 inches on her map stood for 14 feet in her living room.

This meant that every inch on her map stood for 2 feet in the room, which also meant that every ½ inch on her map stood for one foot.

> **Scale: 1/2 inch = 1 foot**
>
> ├─ 1/2 in. ─┤
> **1 foot**

This gave Mia a way to make everything on her map the right size. Easy!

A **map scale** is how the size of something on a map compares to its size in the real world. Mia had just worked out the scale for her map.

4 feet

bookshelf

television

14 feet

chair

rug

3 feet

patio doors

sofa

table

6 feet

5 feet

door

Continuing, she measured the bookshelf. It was 4 feet long. Using her scale of ½ inch = 1 foot, the bookshelf would be ½ inch x 4, which is 2 inches. On her map, she made the bookshelf 2 inches long.

Now Mia could use this scale (1 foot in the living room equals 1/2 inch on the map) to add the other objects in the room to her map. The table, for example, was 5 feet long. This made it 2 ½ inches long on the map.

A scale is written using an = sign. The map size goes first. The scale of Mia's map, for example, was *1/2 inch = 1 foot*. This tells anyone using her map that 1/2 inch on it stands for one foot.

From this they can figure out that 1 inch on her map stands for two feet, 2 inches stands for four feet, and so on.

A map scale is usually written right on the map, often in one corner. Using the scale, anyone can figure out the real measurements or distances being shown on the map.

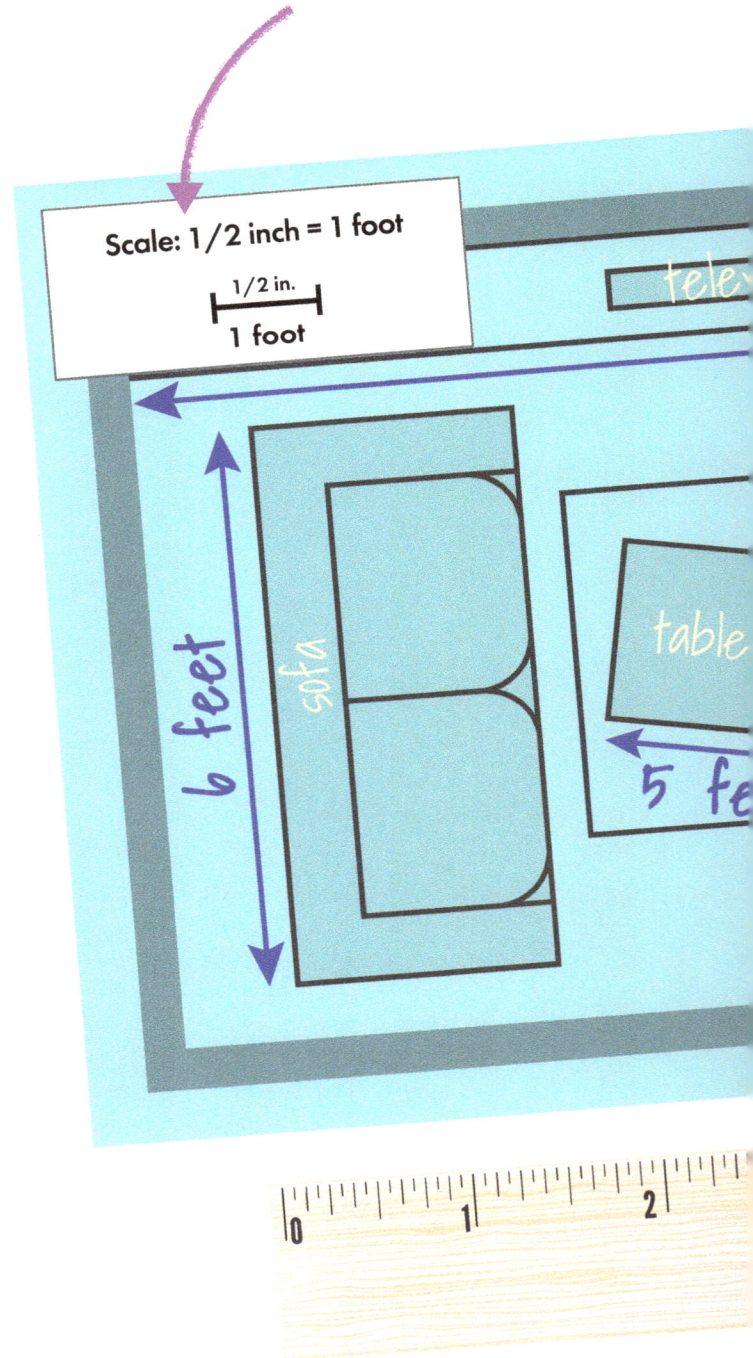

Scale: 1/2 inch = 1 foot

1/2 in.

1 foot

6 feet

sofa

tele

table

5 fe

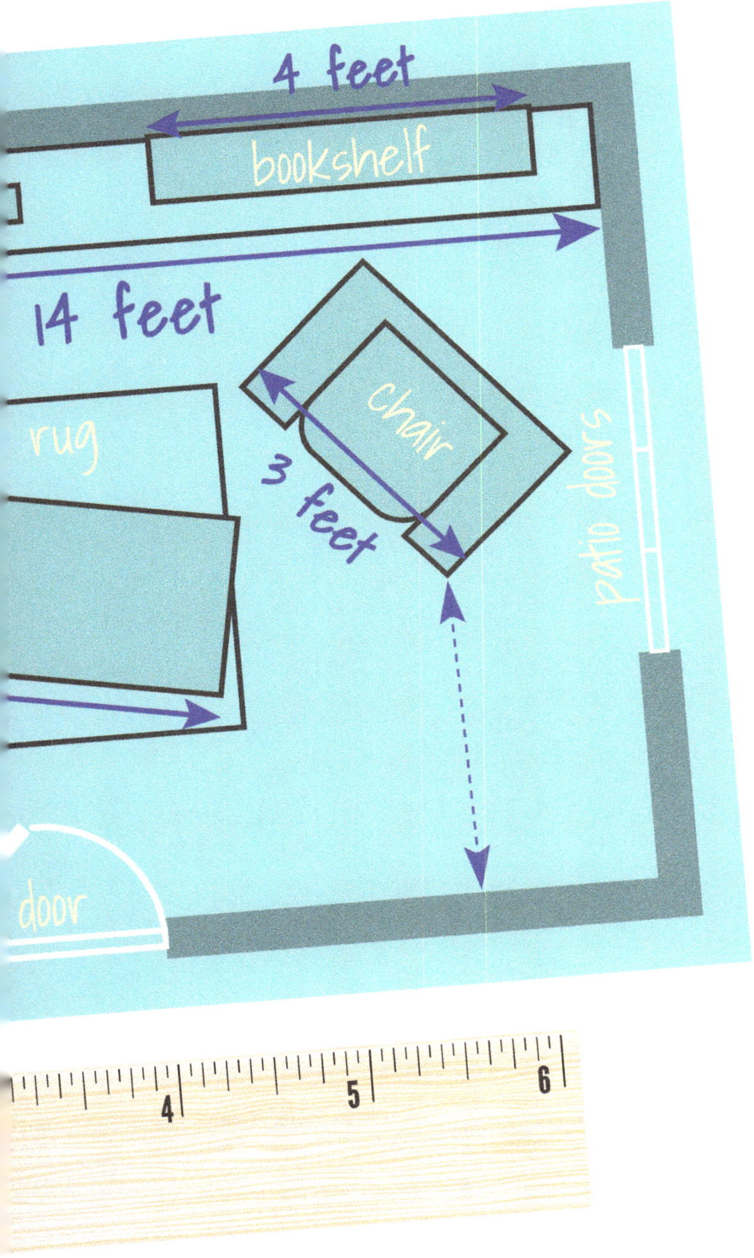
4 feet
bookshelf
14 feet
chair
3 feet
rug
patio doors
door
4
5
6

The scale of a map is useful in two ways.

1. It gives you a way to work out the size of the actual things shown on the map.
2. It gives you a way to figure out how far apart things or places on the map are.

For example, suppose Mia's mom wants to see if another chair about two feet wide would fit between the chair that's there and the wall by the door.

Using the scale of *1/2 inch = 1 foot,* she can measure the distance between them on the map, then figure out how much room there actually is between the chair and the wall.

She finds that the space between the chair and the wall is about 2 inches on the map. Using the scale of *1/2 inch = 1 foot,* she figures out that the distance from the real chair to the wall is about 4 feet. This is enough room for another chair, but not a huge one!

LARGER MAPS

Let's look at a map of something bigger than a living room. Part of a town, for example, is a lot bigger than Mia's living room.

So, on this map, the scale is *1 inch = 1000 feet.* One inch on the map stands for 1000 feet in the real town.

How might you use the map to find out how far it is from the pond to the soccer field?

You could start by measuring the distance on the map. This turns out to be about two and one-half inches. Since the scale is *1 inch = 1000 feet,* the distance from the pond to the soccer field is about 2500 feet.

soccer field

pond

Scale: 1 inch = 1000 feet

1000 feet

EVEN LARGER MAPS

Here is a map of North America with a scale of *1 inch = 1000* miles.

An inch on this map stands for 1000 miles. You could use it to find out how far it is across the United States.

You could use it to figure out how far it is from Canada to Mexico, or how far across the state of Alaska is.

You could use this map to find out the length of the Mississippi River.

So you can use a map scale to figure out distances between places in the real world. You can also use it to measure how large things like states or countries are.

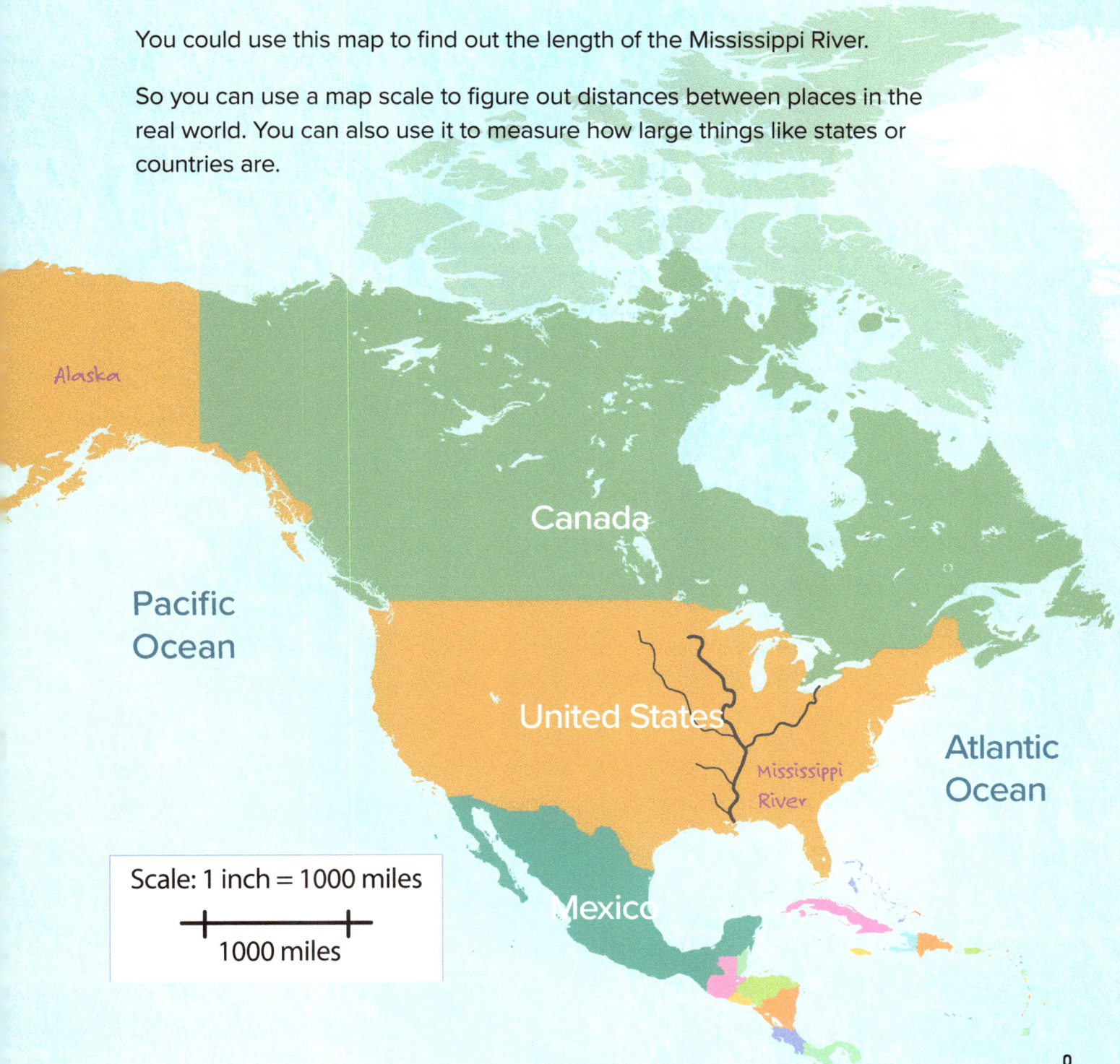

Alaska

Pacific
Ocean

Canada

United States

Mississippi
River

Atlantic
Ocean

Mexico

Scale: 1 inch = 1000 miles

1000 miles

USING A MAP SCALE

For this activity you will need

- a ruler that measures in inches

Steps

① Use a ruler to measure each of these distances on this map to the nearest half-inch.

From the middle of the sofa to the middle of the bookshelf.

- How many half inches is this?
- How far is this in feet?

From the middle of the table to the middle of the television.

- How many half inches is this?
- How far is this in feet?

From the middle of the sofa to the middle of the television.

- How many half inches is this?
- How far is this in feet?

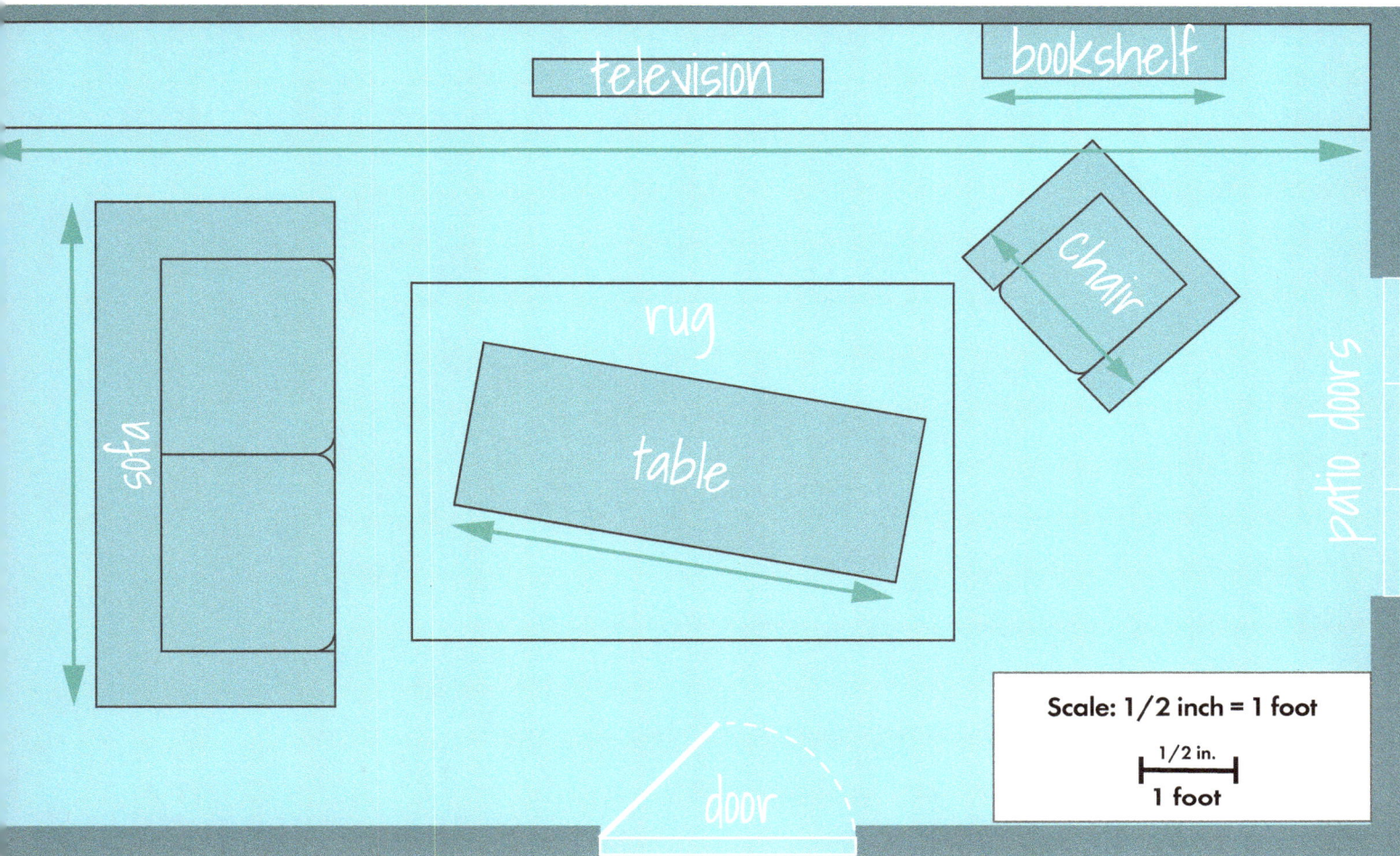

Scale: 1/2 inch = 1 foot

1/2 in.
1 foot

2 Suppose you were making a map of your living room and you wanted to have a scale of *1 inch = 1 yard.*

- If there were a bookshelf one yard wide, how wide would it be on your map?

- If there were a table three yards long, how long would it be on your map?

- If there were a rug 14 feet long, about how many inches long would it be on your map?

TOWN MAP

For this activity you will need

- a ruler that measures in inches

Steps

1. Use a ruler to measure each of these distances on the map to the nearest inch.

   From the South Woods to the school.

   - How many inches is this?
   - How far is this in feet?

   From Jane's farm to the store.

   - How many inches is this?
   - How far is this in feet?

   From the apple orchard to the library.

   - How many inches is this?
   - How far is this in feet?

2. Measure something else on the map that you choose.

   - How many inches is it?
   - How far is this in feet?

Scale: 1 inch = 1000 feet

1000 feet

North Woods

apple orchard

Blue River

Jane's Farm

South Woods

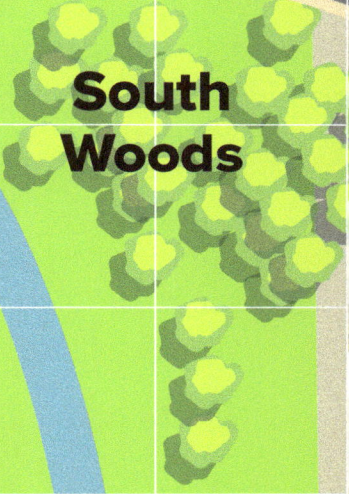

gas

theater

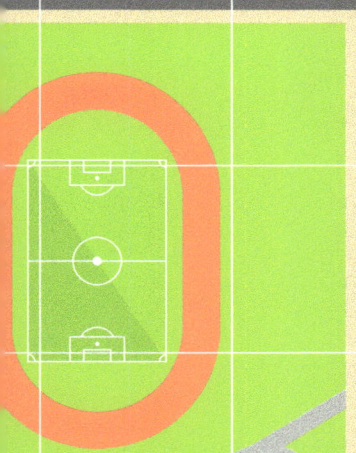

store

bank **library**

school

wood mill

Let's Do This!

MAKE A MAP

For this activity you will need

- a sheet of paper

- a yardstick or tape measure

Steps

In this activity you will use a scale of *1/2 inch = 1 foot* to make a map of the room you are in. (If the room you're in is very large, you might choose to use a scale of 1/4 inch = 1 foot.)

1 Measure the room to find out how long it is from wall to wall.

2 Measure to find out how wide it is from wall to wall.

3 Measure how far your chair is from the front wall of the room; how far from the left wall of the room.

4 Use the scale of *1/2 inch = 1 foot* (or *1/4 inch = 1 foot*) to draw the walls of the room and to show your chair in the right place.

5 Put some other objects in the room on your map.

Save your map. You'll need it later on!

3 CHAPTER SYMBOLS

Now that we know how to use map scales, a new question comes up. What about things like trees, mountains, rivers, lakes, roads and towns? How are these shown on maps? They are often shown using symbols.

Let's see how this works on our town map.

Instead of writing "school," "library" and "store" on the map, we can use a symbol to stand for each of these.

 school

 library

 store

Now it looks like this.

Much simpler!

A **symbol** is a mark, a shape, or even a small picture that stands for something. A symbol gives someone a certain idea without using any words.

Too much writing on a map can make it hard to read. This is why many maps use symbols. When you know what the symbols mean, the map is a lot easier to use.

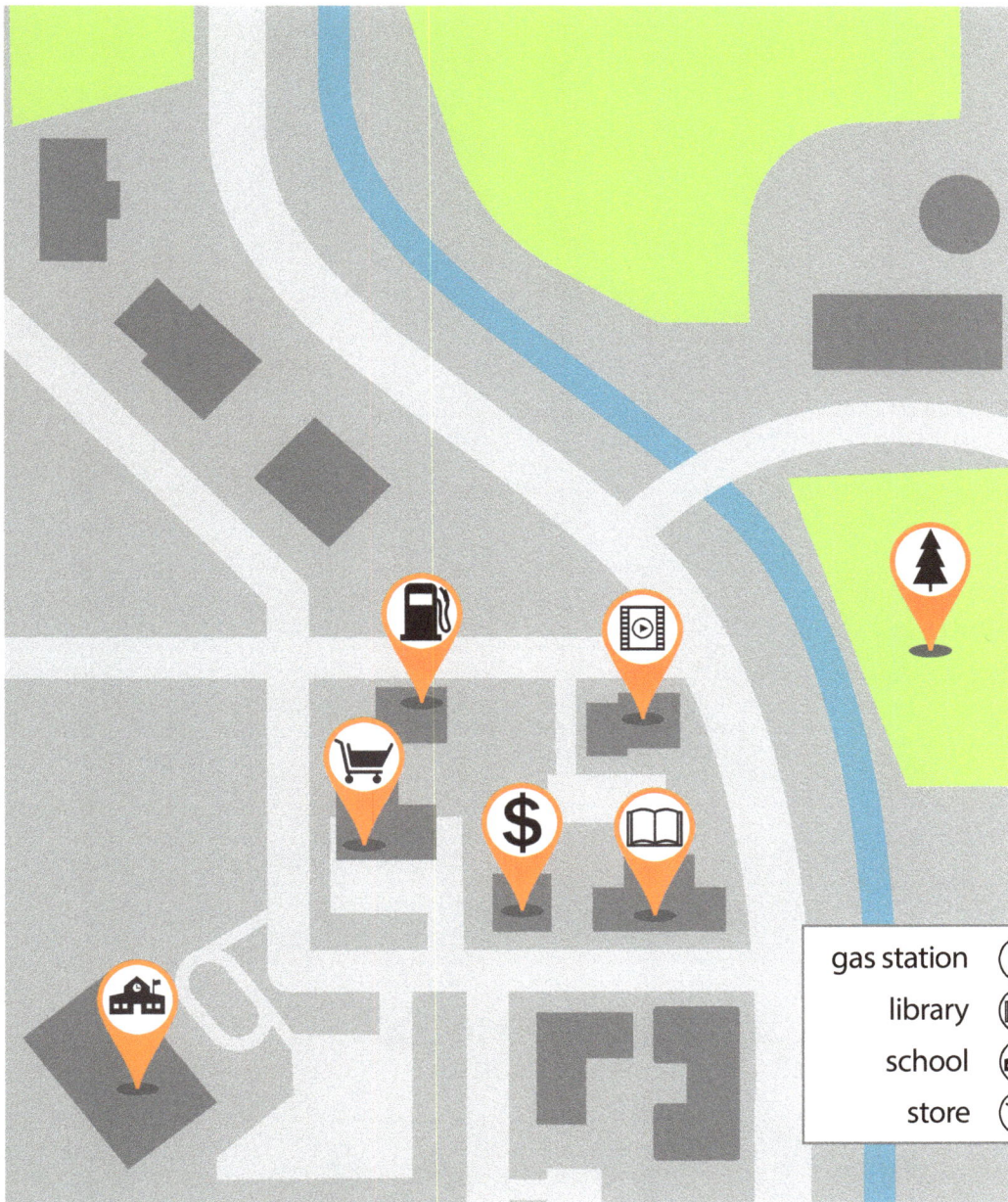

We can make the map even simpler by using more symbols.

The roads and the river don't need symbols because it's easy to tell what they are.

Now the map has all the same information as before, but it's a lot easier to see where everything is.

gas station		bank	
library		theater	
school		South Woods	
store			

THE MAP KEY

To be useful, any map that has symbols on it should include an explanation of what the symbols stand for. This is done with what we call the map "key."

The **key** lists the symbols used on a map along with what each one stands for. It opens the door to understanding what's on the map.

KEY

gas station	⛽	bank	$
library	📖	theater	📽
school	🏫	South Woods	🌲
store	🛒		

MORE MAP SYMBOLS

Some maps use other symbols as well.

★ A capital is a city where government officials of a state or country meet. On many maps the capital of a state is shown with a star.

⍟ The capital of a country is usually shown using a star with a circle around it.

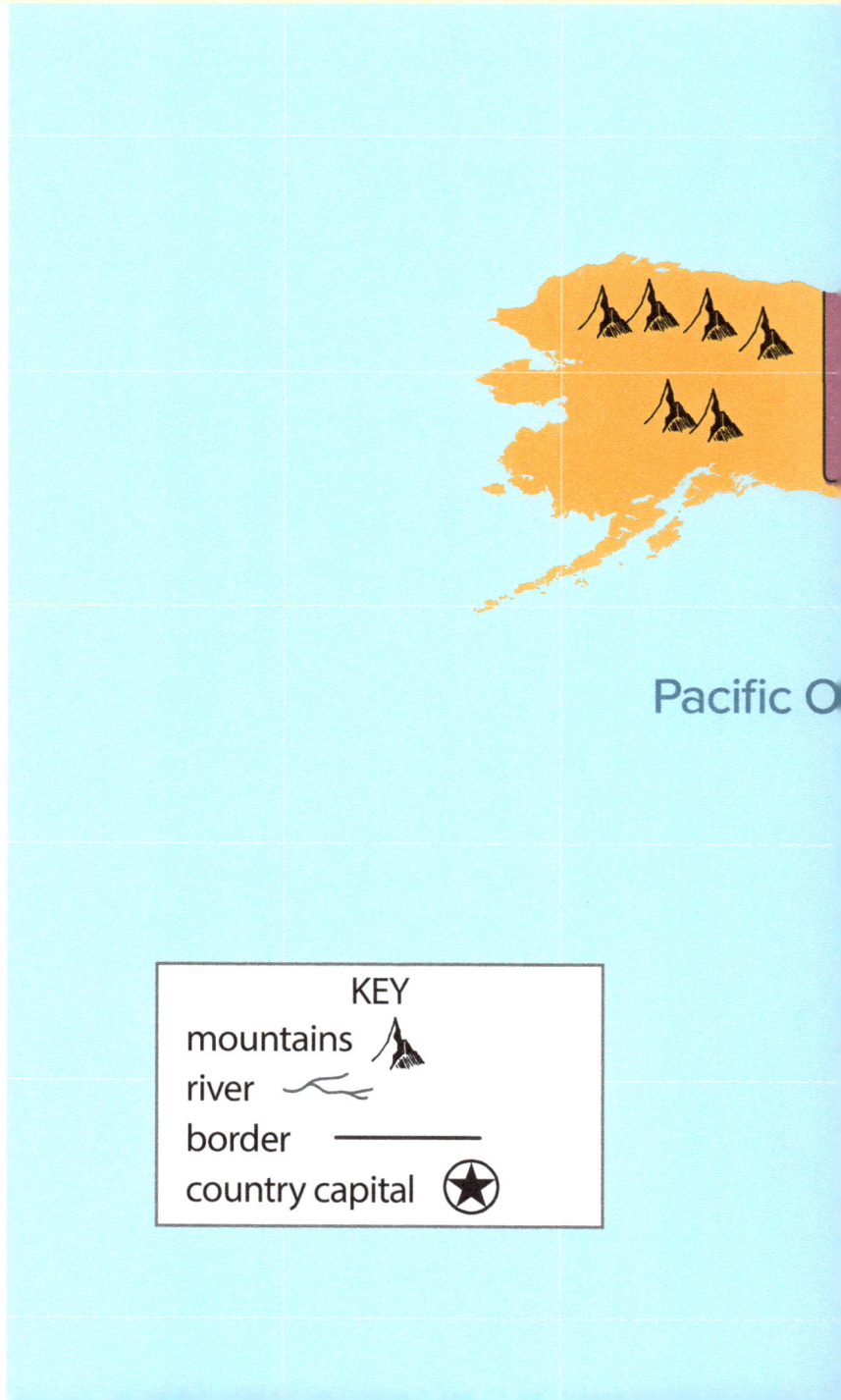

Pacific O

KEY

mountains	⛰
river	～
border	——
country capital	⍟

Many maps show lines between states or countries. These are called **borders.** Borders between states are usually shown with dotted lines. Borders between countries are usually shown with a thick gray or black line.

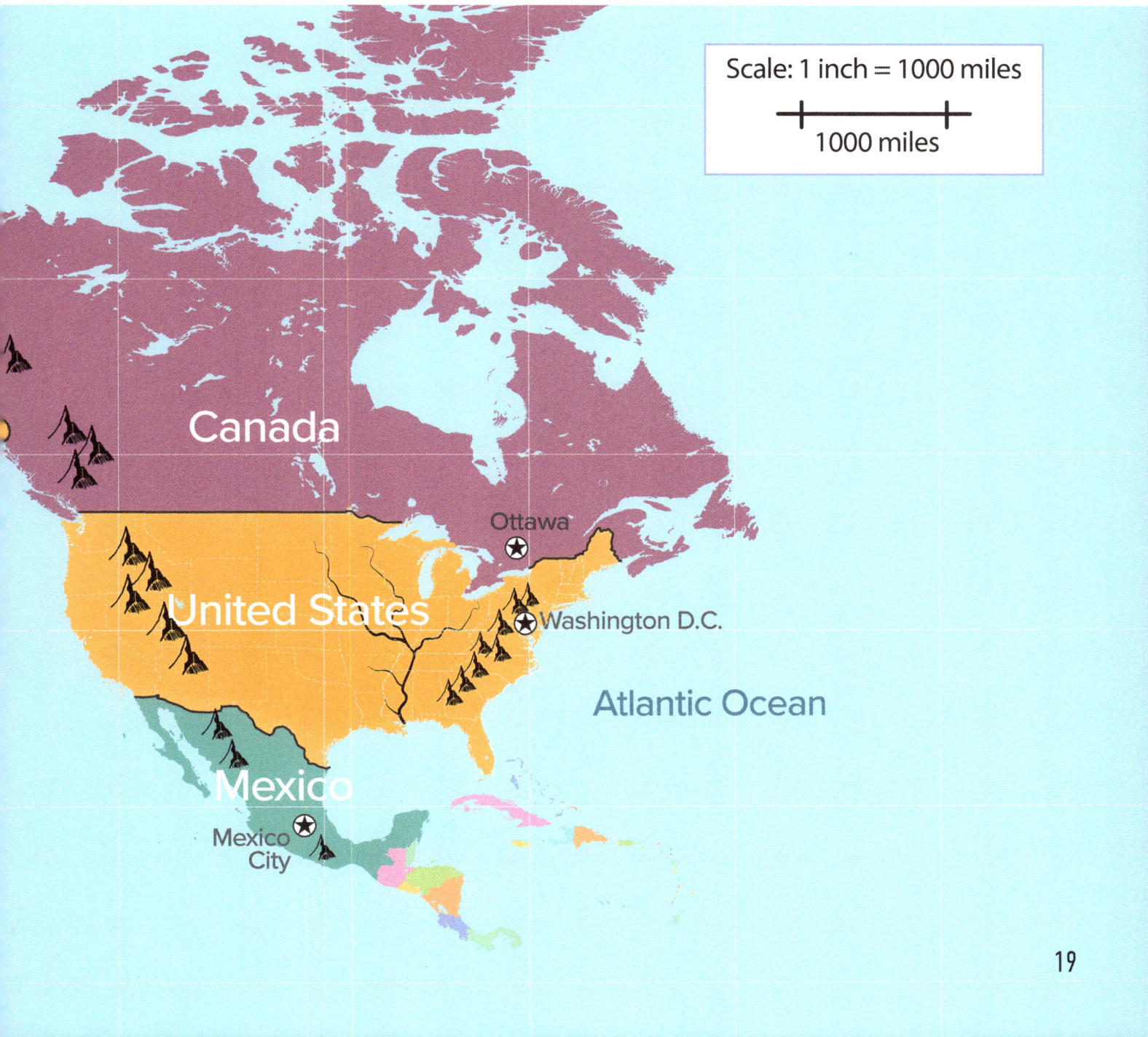

Scale: 1 inch = 1000 miles

1000 miles

Canada

Ottawa

United States

Washington D.C.

Atlantic Ocean

Mexico

Mexico City

Not all maps have a key. But when they do, it is usually found in a box near one corner of the map. A map key is sometimes called a **legend**.

Take a look at this map and key. What are some of the things you can learn from this map?

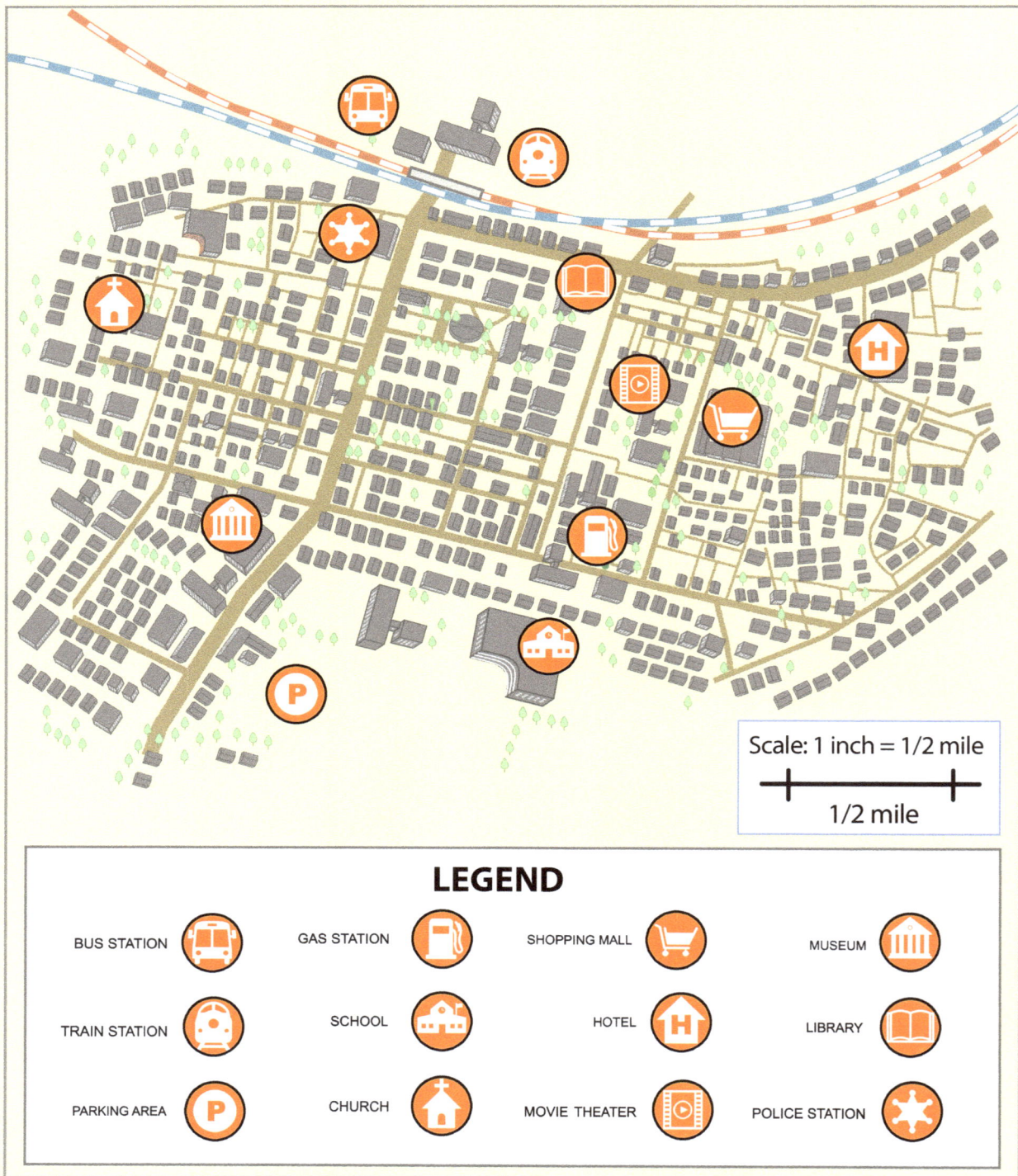

Scale: 1 inch = 1/2 mile

1/2 mile

LEGEND

BUS STATION	🚌	GAS STATION	⛽	SHOPPING MALL	🛒	MUSEUM	🏛
TRAIN STATION	🚆	SCHOOL	🏫	HOTEL	H	LIBRARY	📖
PARKING AREA	P	CHURCH	⛪	MOVIE THEATER	🎬	POLICE STATION	✦

When you're using a map, the key, or legend, can help you understand what the different symbols on it stand for.

EXPLORING A MAP

Let's Do This!

For this activity you will need

- a map of your city or state

- a ruler that measures in inches and centimeters

- another person to work with

Steps

1. On the map of your city or state, find the key.

2. Find the scale and see if it is in inches or centimeters. Check which it is with a ruler if you want to.

3. On the map, find three of the symbols given on the key.

4. Choose two places on the map and measure the distance between them to the nearest inch or centimeter. Using the scale, figure out about how far it is between the two places.

5. Show another person the roads you would take to travel between the two places you chose and tell how much distance this would be. (It might be different from the distance in the last step.)

4 CHAPTER NORTH, SOUTH, EAST AND WEST

To use a map to get from one place to another, you need to know something about directions on our planet and how they work.

On a map, a **direction** is the way something faces, the way it's pointing, or going.

Tom is pointing in one direction. Erica is pointing in another direction. Rufus is facing another direction.

When you're going from one place to another, you're always going in a direction. And when you use a map, you're figuring out what direction to go.

To use a map, you need to know a bit about how directions work.

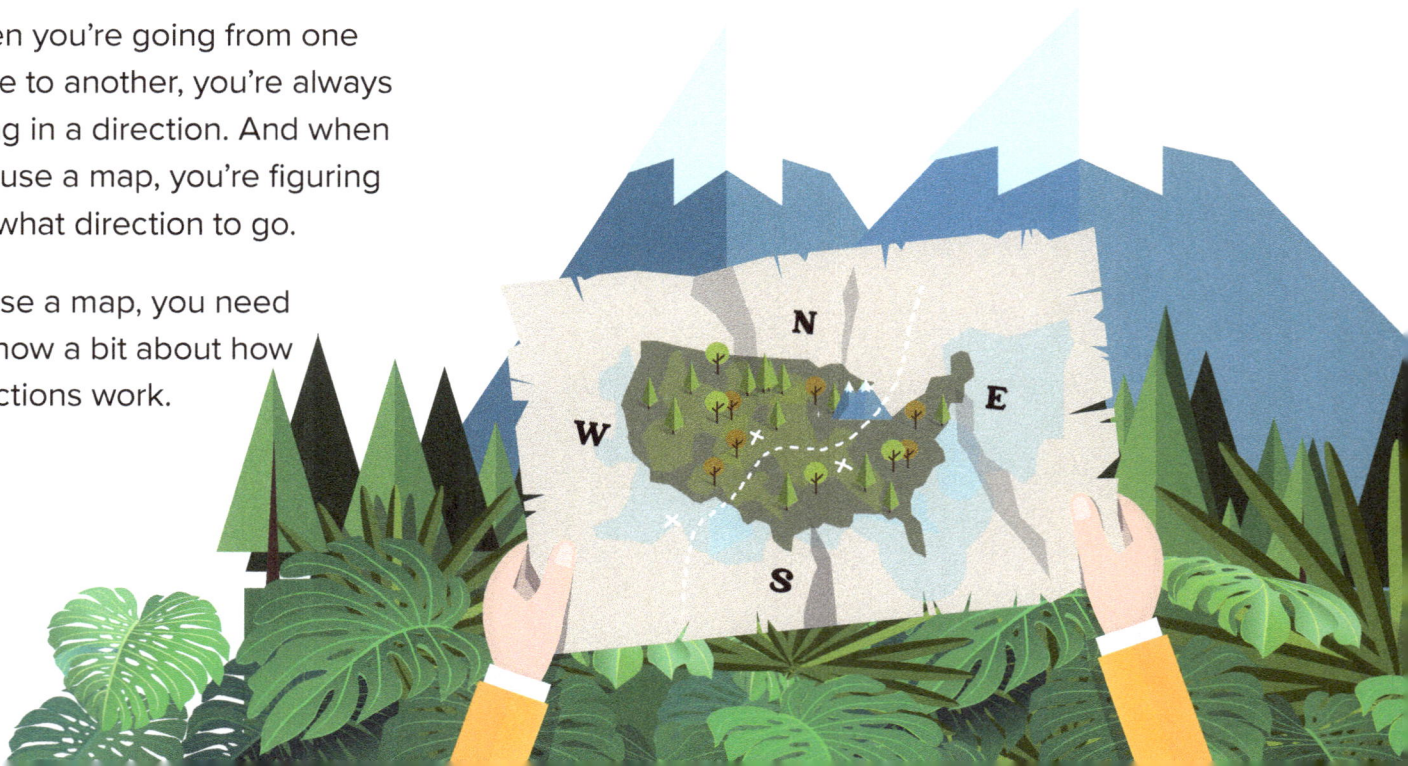

THE MAIN DIRECTIONS

From anywhere you are on Earth, there are four main directions you can go in. You can go toward the top of the planet, toward the bottom, or toward one side or the other. These are the four main directions, called north, south, east and west.

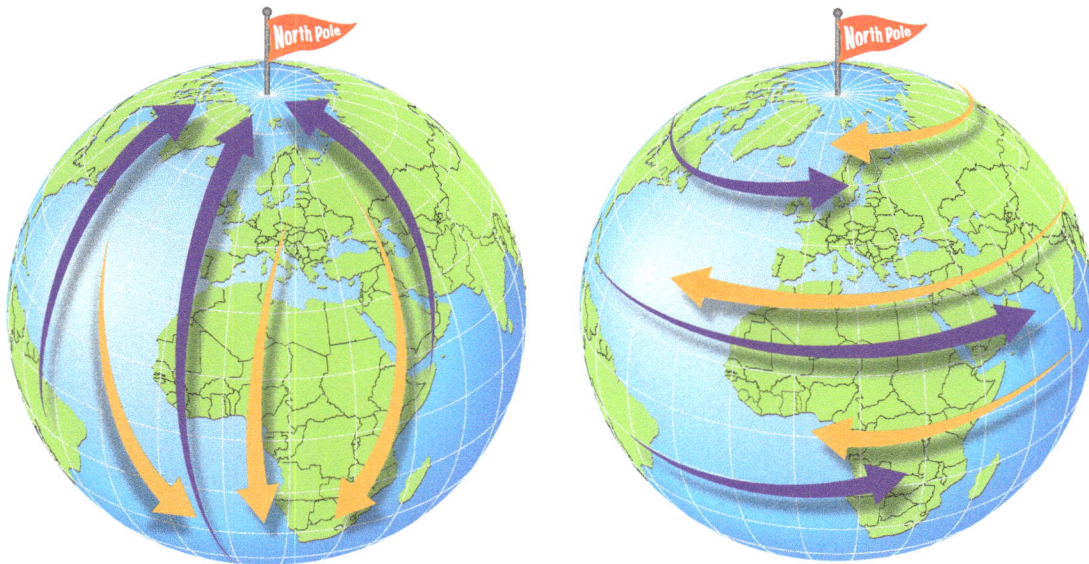

North is the direction toward the top of the planet, which we call the **North Pole**. Wherever you are on the earth, if you're headed toward the North Pole you're going north.

It's not always easy to tell where the North Pole is. To do that we use a compass.

A **compass** is a small, round device with a needle that always points to the North Pole.

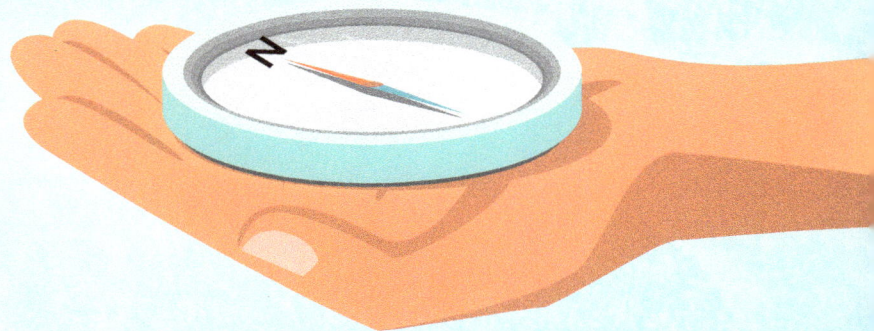

You can always find which way is north by holding a compass flat in your hand and seeing which way the needle points. That's north!

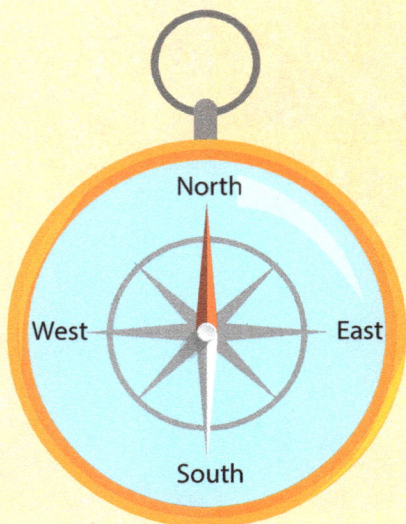

If you were to go in the opposite direction, you'd be headed south. The farthest south you can go is the **South Pole** at the very bottom of the world.

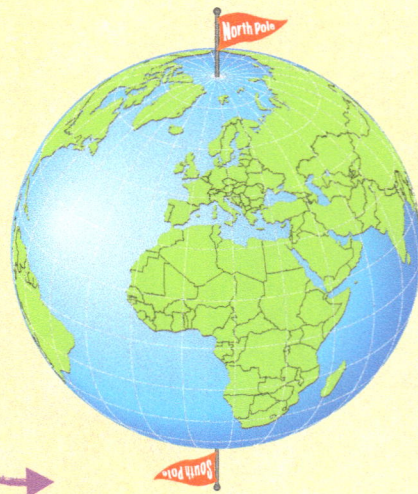

When you face north, **west** is the direction to your left. The sun sets in the west.

When you face north, **east** is the direction to your right. The sun rises in the east.

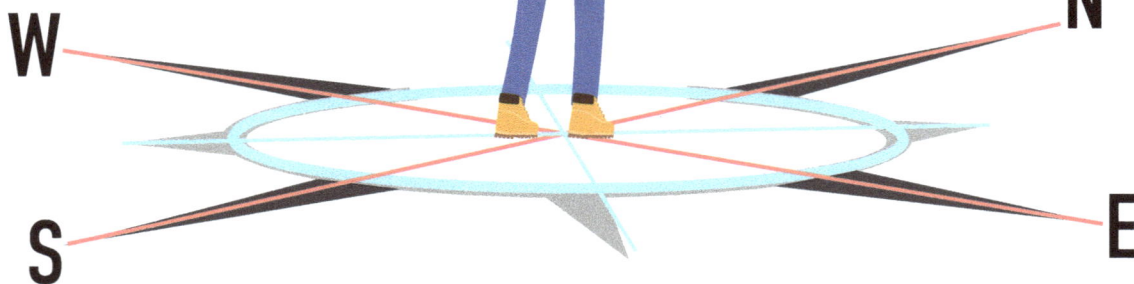

OTHER DIRECTIONS

Knowing the four main directions is very useful. Sometimes though, when using directions, you need to be more exact.

So there are four more directions that lie in between the four main directions we've already talked about.

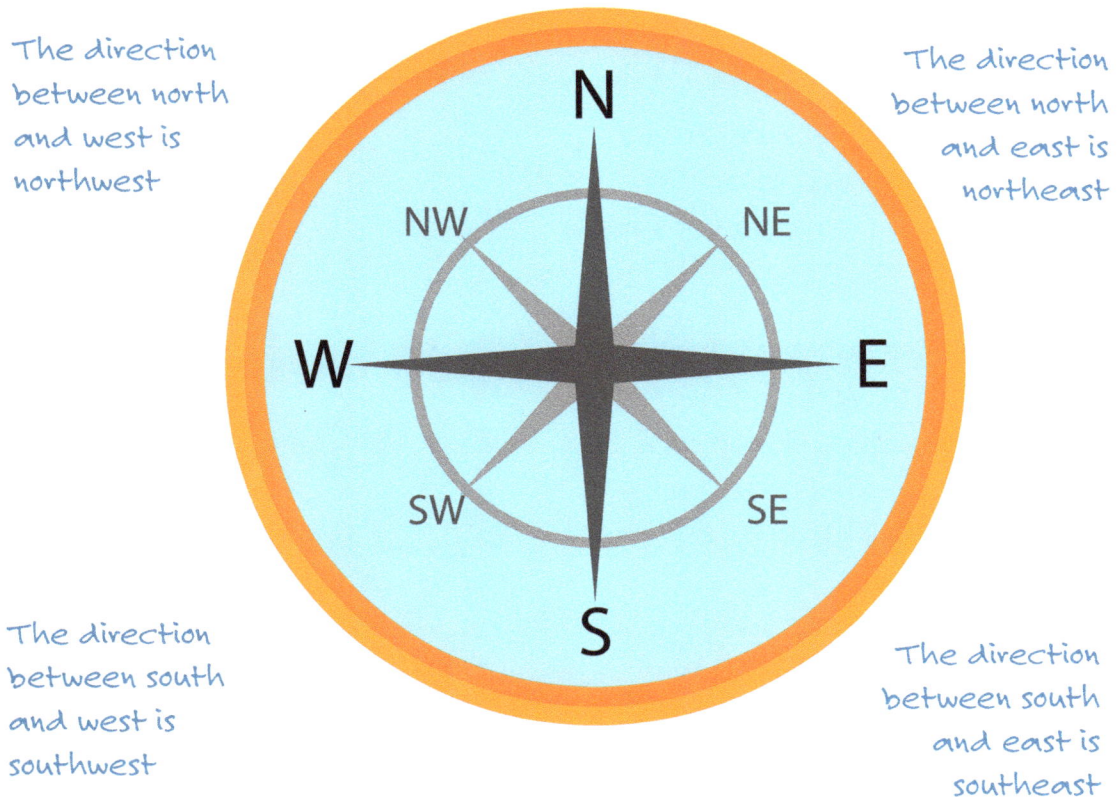

The direction between north and west is northwest

The direction between north and east is northeast

The direction between south and west is southwest

The direction between south and east is southeast

Often, the names of all eight directions are shortened to make them easier to write.

north is	N	northeast is	NE
south is	S	southeast is	SE
east is	E	northwest is	NW
west is	W	southwest is	SW

Using these eight directions, you can tell someone more exactly what direction to go in.

USING DIRECTIONS ON A MAP

Let's see how this works with our town map.

See the star shape in the bottom left corner of the map? You will find a symbol like this on most maps.

Because it looks something like a flower, it's called a **compass rose**.

The compass rose shows you the directions on the map. It tells which way is north, which is south, and so on. On this map only the four main directions are labeled.

Using a compass rose you can tell which way on a map is north. When you're standing at a certain place shown on a map, you can also tell what direction to go to get to some other place on the map.

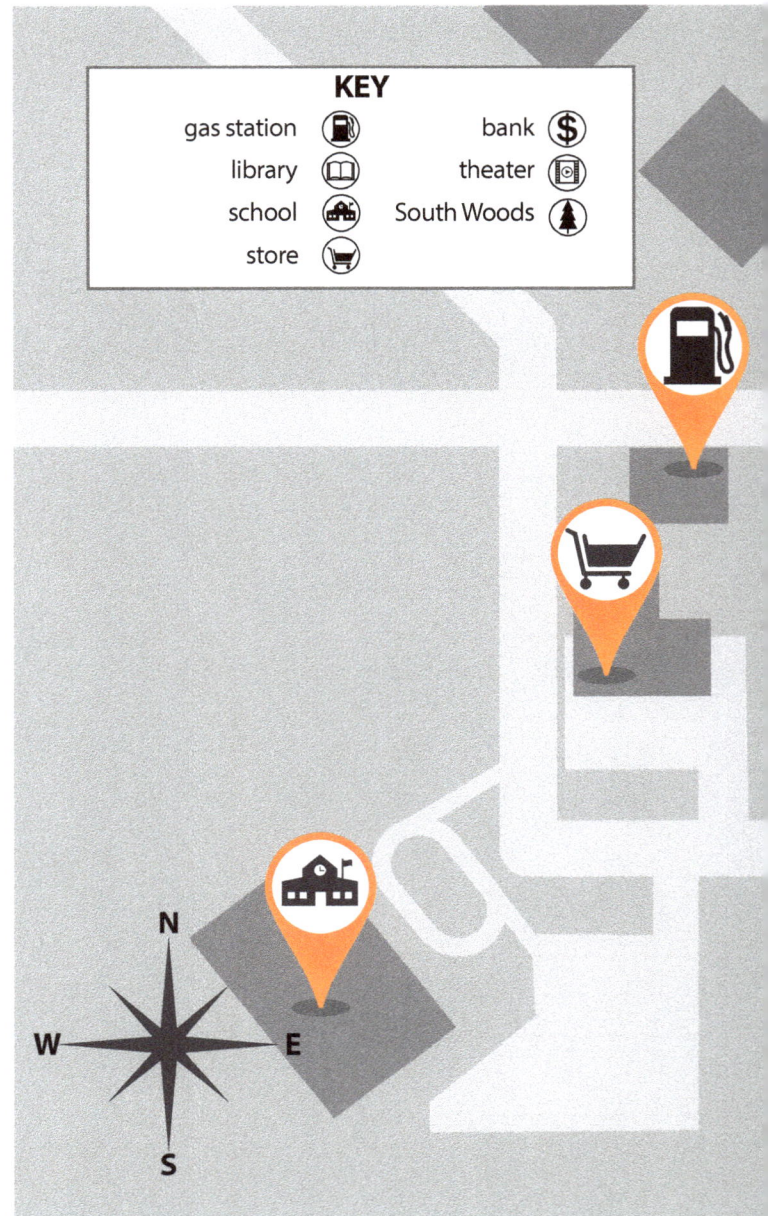

KEY

gas station	🛢	bank	$
library	📖	theater	🎬
school	🏫	South Woods	🌲
store	🛒		

ONE MORE THING

Sometimes, instead of a compass rose, north on a map is shown with just a small north-pointing arrow marked N.

If north isn't shown with an arrow or compass rose, the top of the map is north.

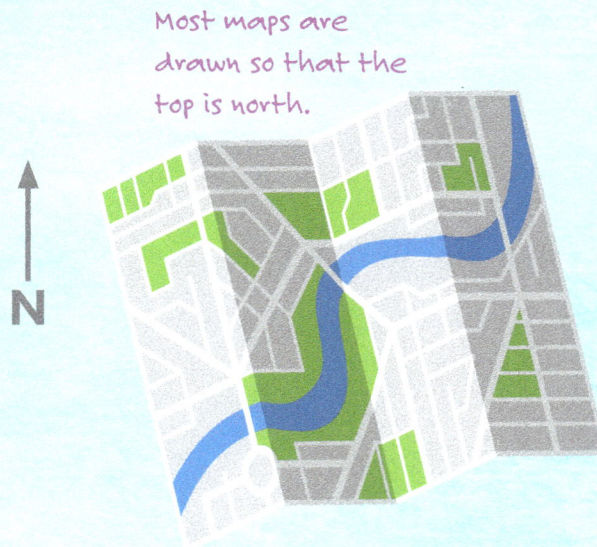

Most maps are drawn so that the top is north.

Scale: 1 inch = 1000 feet

1000 feet

For example, on our town map you can see that if you start at the library and go southwest, you will get to the school.

If you stand at the school and look northeast, you will see the store.

You could say the store is northeast of the school. Or you could say the school is southwest of the store. Both are true!

CITY ZOO

When you're using a map to tell you which way to go, it helps to turn it so that north on the map is pointing north. This way you'll be able to head in the right direction.

You can use a compass to do this.

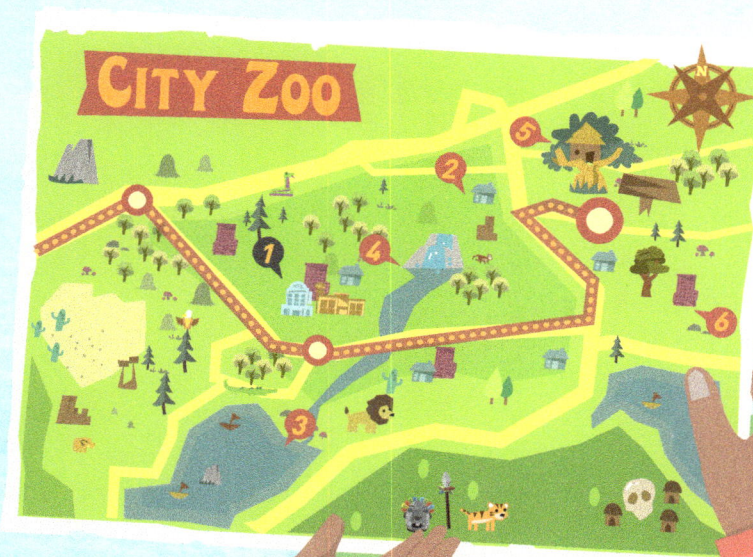

USE A COMPASS

For this activity you will need

- a compass
- a partner to work with

Steps

Do the following steps at two different places inside, and then three different places outside. Choose places that are far apart. Also choose places where there is no metal close by, because the compass needle will point to that instead of the North Pole.

1 Hold a compass flat in your hand and wait for the needle to stop moving. It may swing back and forth a bit but will soon stop and point north.

2 Face the direction the needle is pointing. This is north. Turn the compass so that the N (for north) matches the way you are facing and the needle is pointing.

3 With your hand, point north and tell your partner some things that are north from where you are.

4. Now turn in the opposite direction from north. This is south. With your hand, point south and tell your partner some things that are south from where you are.

5. Now face north again. With your hand, point east and tell your partner some things you see that are east from where you are.

6. With your hand, point west and tell your partner some things you see that are west from where you are.

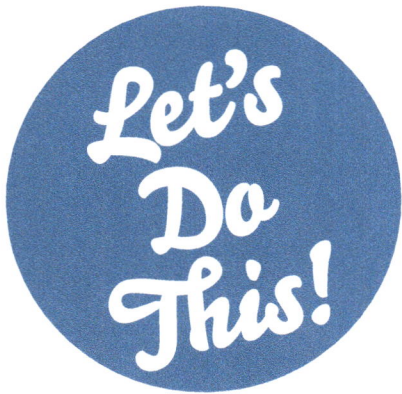

WHICH WAY DO I GO?

Steps

Use this map to answer the questions.

1. If you were in Sheridan, which direction would you have to go to get to Amity?

2. If you were in Salem, which way would you have to go to get to Dallas?

3. If you were in McMinnville, which way would you have to go to get to Amity?

4. Which direction is Dallas from Willamina?

5. Which direction is McMinnville from Sheridan?

6. Which direction is Sheridan from Salem?

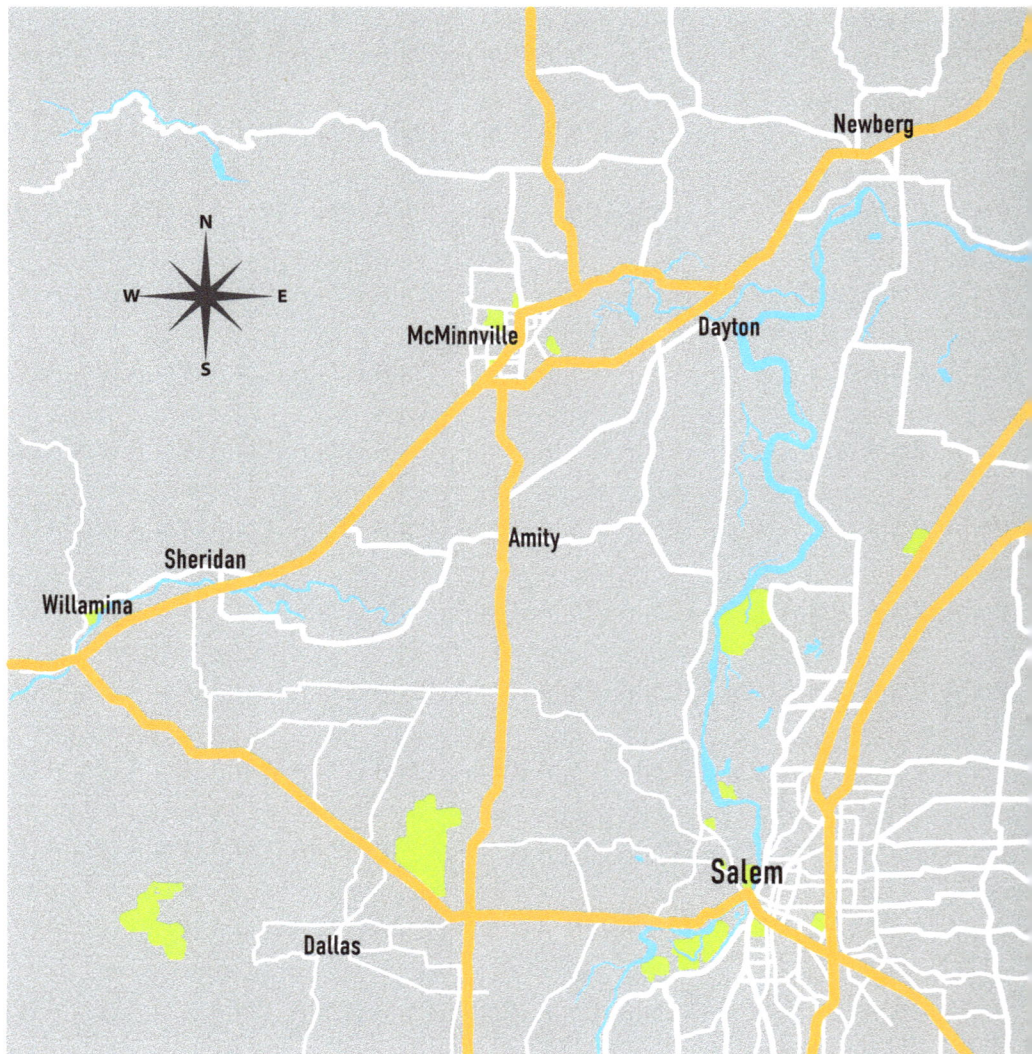

COMPASS ROSE

Let's Do This!

For this activity you will need

- a compass
- the map you made in the Make a Map activity

Steps

1 Use the compass to find out which way is north in the room you used for your map.

2 Draw a compass rose on your map, with the north arrow pointing in the right direction. Show all eight directions on your compass rose.

3 Save your map to use later.

5 CHAPTER MAP GRIDS

Back when people were first beginning to make and use maps, a smart mapmaker figured out a way to find something on a map quickly and easily.

Let's see what this new idea was and how it worked.

Maps often have lots of information on them. Some maps show many streets, highways, rivers, parks, airports and sometimes even more. This can make the map seem complicated. It can make it hard to find exactly what you're looking for.

To help with this, many maps have lines marked on them that form a pattern of equal-sized squares. It looks like this:

This is called a map **grid**.

The squares form what we call rows and columns.

On a map grid, the rows are usually labeled with letters.

Then the columns are labeled with numbers.

Using these, you can name any square on a map just by giving its letter and number.

Let's see how this works on a real map.

Here is our town map. If you find the **I-2** square, you'll see that the school is in it.

The gas station is in **E-5**.

In this grid there is an x in the square that is formed where row B and column 3 meet. We say the x is in square B3.

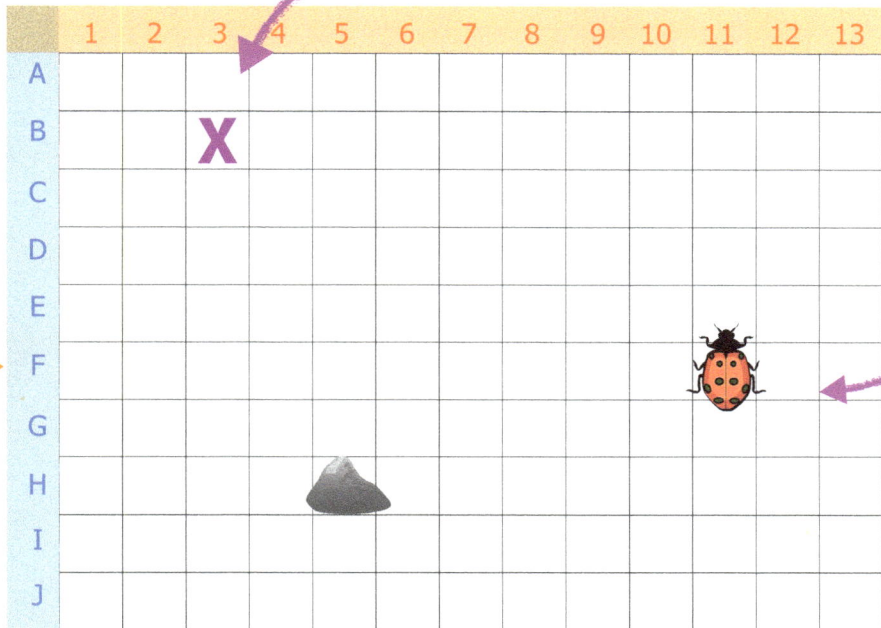

The ladybug is in square F11.

The rock is in square H5.

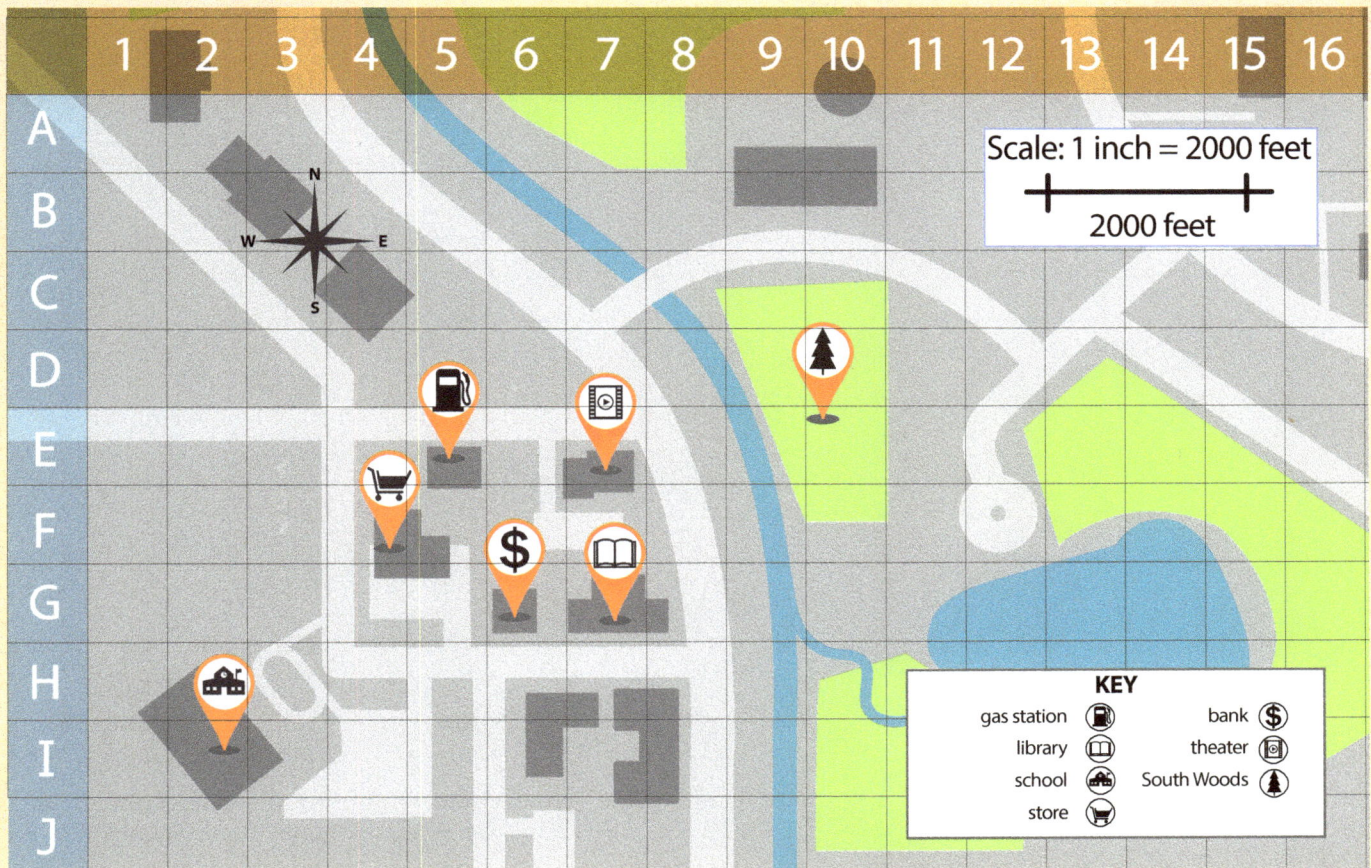

	1	2	3	4	5	6	7	8	9	10	11	12	13
A													
B			X										
C													
D													
E													
F											🐞		
G													
H					🪨								
I													
J													

Scale: 1 inch = 2000 feet
2000 feet

KEY

gas station	⛽	bank	$
library	📖	theater	🎬
school	🏫	South Woods	🌲
store	🛒		

Using the grid on a map makes it much easier to find what you're looking for.

A map with a grid like this map of Connecticut often has a list of places on it. These are all places on the map, and each one is followed by the letter and number of the square in which it can be found.

You can use the list to quickly locate a place on the map, using its letter and number to find the right square. There, it will be easy to find.

	1	2	3	4	5	6	7
A							
B							
C							
D							
E							
F							
G							
H				New Milford			
I							
J							
K			Danbury	84	Newto		
L							
M							
N							
O	684		Norwalk	Westport	Fa		
P		Stamford	95				
Q	Greenwich						
R							

Branford	M13	Enfield	C16
Bridgeport	O7	Fairfield	O7
Bristol	G11	Glastonbury	G16
Cheshire	I11	Greenwich	Q2
Danbury	K4	Groton	L23
East Hartford	F15	Hamden	K12
East Haven	M12	Hartford	E15

Enfield

Putnam

91 **84** **395**

Windsor

Vernon

Mansfield

291

West Hartford

HARTFORD

Manchester

East Hartford

384

Newington

Glastonbury

Bristol

84

New Britain

Southington

Waterbury

691

Meriden

Middletown

Cheshire

Norwich

Naugatuck

Wallingford

395

Hamden

91

New London

Groton

95

New Haven

West Haven

East Haven

Branford

95

Milford

Stratford
port

n

l

Scale: 1 inch = 15 miles

15 miles

Manchester	E17	New Haven	L11	Putnam	C25	Vernon	D17
Mansfield	E20	New London	L22	Shelton	L9	Wallingford	J13
Meriden	I13	New Milford	H5	Southington	H12	Waterbury	I10
Middletown	I15	Newington	F14	Stamford	P3	West Hartford	E14
Milford	N9	Newtown	K7	Stratford	N8	West Haven	M11
Naugatuck	J10	Norwalk	O5	Torrington	E9	Westport	O5
New Britain	G13	Norwich	I23	Trumbull	M7	Windsor	D15

35

Let's Do This!

FIND IT FAST

For this activity you will need

- a map of your state

- a partner to work with

Steps

1. Have your partner point out places on the map. You tell what grid squares these places are in. Do this until you can do it easily.

2. Have your partner use the list of places and grid locations on the map to tell you the name of a place and what grid square it is in. Then you find the place. Do this until you can do it easily.

ADD A GRID

Let's Do This!

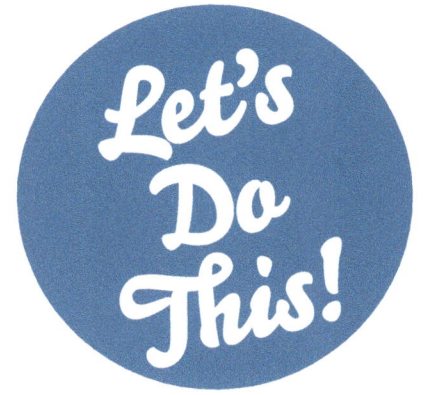

For this activity you will need

- the map you made in the Make a Map and Compass Rose activities

- a ruler or yardstick

- pencil

Steps

1 Use your yardstick or ruler to draw a grid on your map.
   Make your lines one or two inches apart.

2 Along the top or bottom of the map, number each column of the grid.

3 Along one side, put a letter in each row.

4 On the back of your map, make a list of the things on your map with
   the letter and number of the grid square each is in.

6 CHAPTER A DIFFERENT GRID

People often use maps to show things like cities, states and countries. Maps like these can give you a good picture of a particular area.

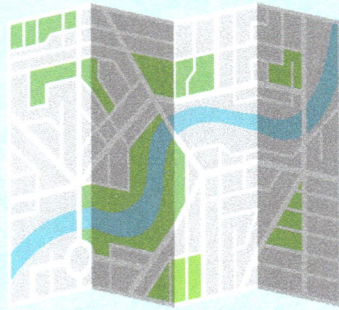

But to learn more about where things are on the earth, people often use globes.

A globe, being shaped like the earth, gives us a clearer idea of the whole planet and where things are on it.

GRID LINES

When looking at a flat map, you can use its grid to find something. When you want to find something on a globe, you also use a grid but it works a little differently.

When you look at a globe, you will see a set of lines running from top to bottom, from the North Pole to South Pole. You see another set of lines that go around the globe. These two sets of lines make a grid.

Keep in mind that these lines are imaginary. They're not actually part of the earth. But they are part of a globe, and can be very helpful when using a globe to find places on the earth.

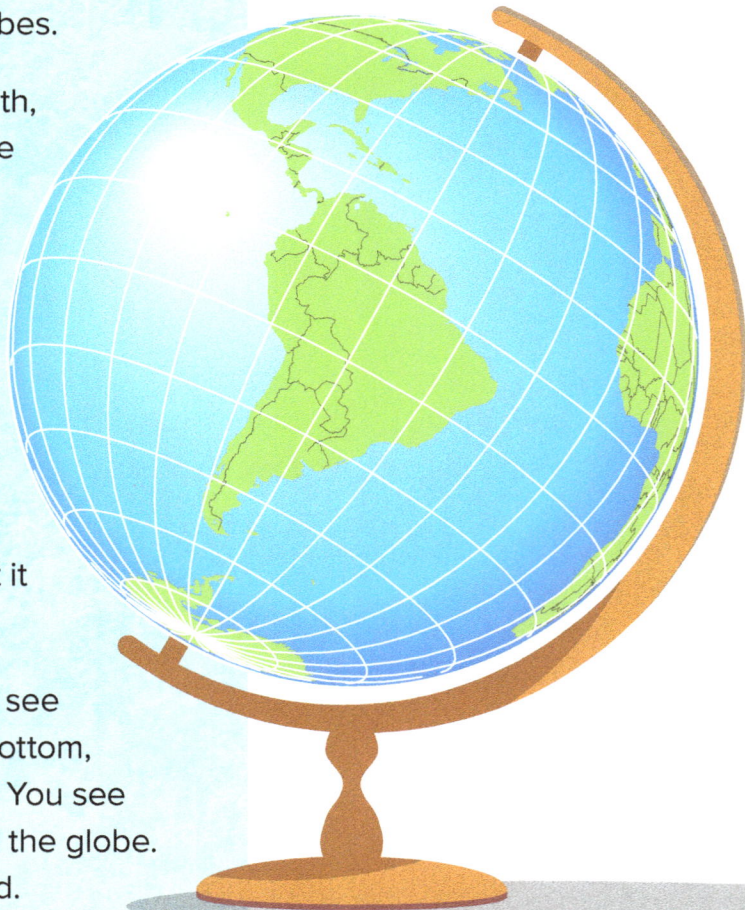

LATITUDE LINES

The lines that go *around* the globe are called **latitude lines**. The longest latitude line is the **equator**. The equator is the imaginary line around the middle of the earth that separates it into two equal parts.

Look at a globe. Starting at the equator, go north toward the North Pole, noticing each latitude line. Do the same going south toward the South Pole.

We use these latitude lines to tell how far north or south something is on the globe.

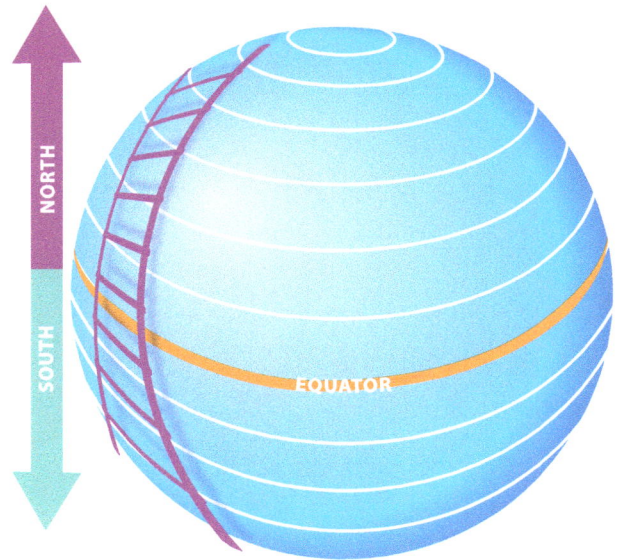

LONGITUDE LINES

Lines that run up and down the globe, from the North Pole to the South Pole, are called **longitude lines**.

Looking at a globe again, find a longitude line. Starting there, go around the earth noticing each longitude line.

We use these longitude lines to tell how far east or west something is on the globe.

The longitude and latitude lines together make a grid!

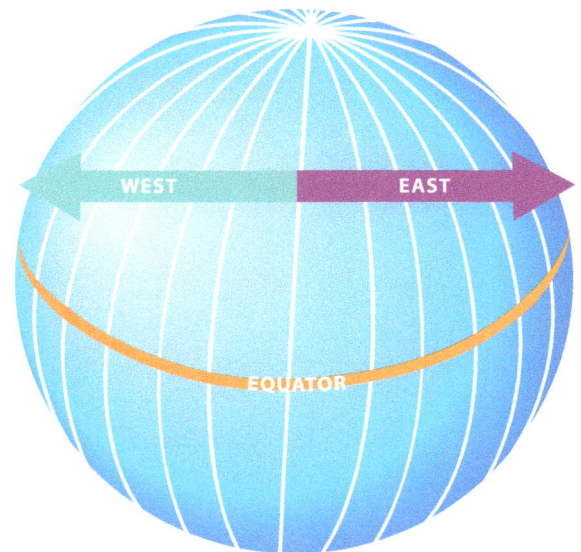

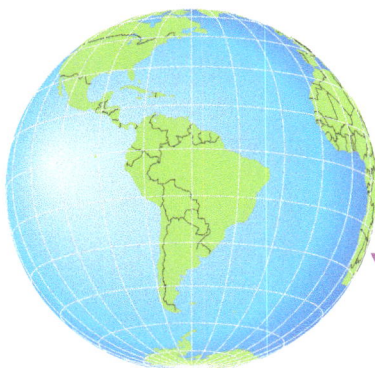

When we see the latitude and longitude lines together on a globe, they look like this.

7 CHAPTER HOW A GLOBE WORKS

The grid of latitude and longitude lines makes it much easier to find any place on a globe.

Latitude and longitude lines are numbered, and they're numbered in what we call **degrees**. These are not the same degrees we use to measure temperature. If you can imagine dividing a circle into 360 equal parts, each part is a degree. A degree is one of the 360 equal parts of a circle.

A special symbol for degree is °. So 360 degrees is written 360°.

If you go 1/4 of the way around a circle, you reach 90°. If you go 1/2 way, you reach 180°. All the way around comes to 360°.

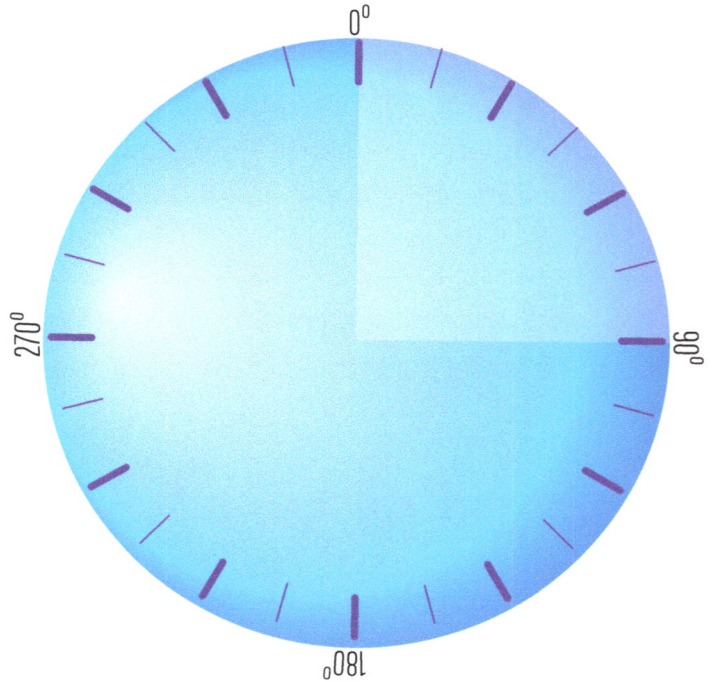

DEGREES FOR LATITUDE LINES

We think of the equator as the starting point for latitude lines, so we call it 0°. From there, the imaginary latitude lines circle the earth, 1° apart, from the equator to the North Pole, and from the equator to the South Pole.

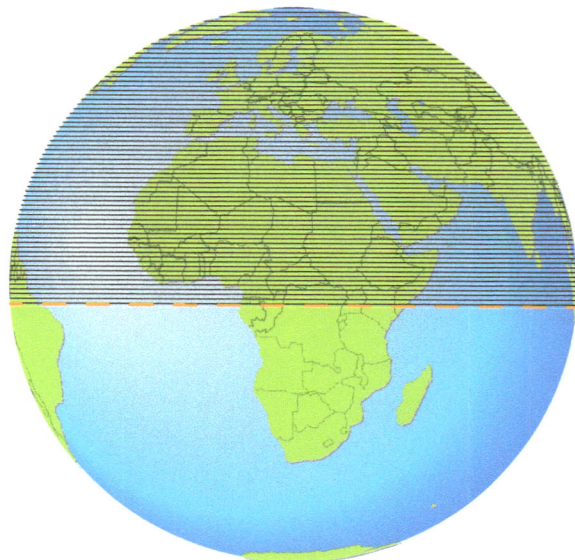

There are 90 latitude lines from the equator to the North Pole and 90 from the equator to the South Pole. That's a total of 180 imaginary lines circling the earth.

There isn't enough room on a globe to write in all 180 latitude lines with their numbers. So most globes show the latitude lines for every 15°, starting at the equator and running to the North Pole. It's the same running from the equator to the South Pole.

Latitude lines *north* of the equator are marked with N for north. The lines to the south of the equator are marked with S for south.

For example, 45° S is 45 degrees south of the equator. 45° N is 45 degrees north of the equator. You can probably find these lines on a globe.

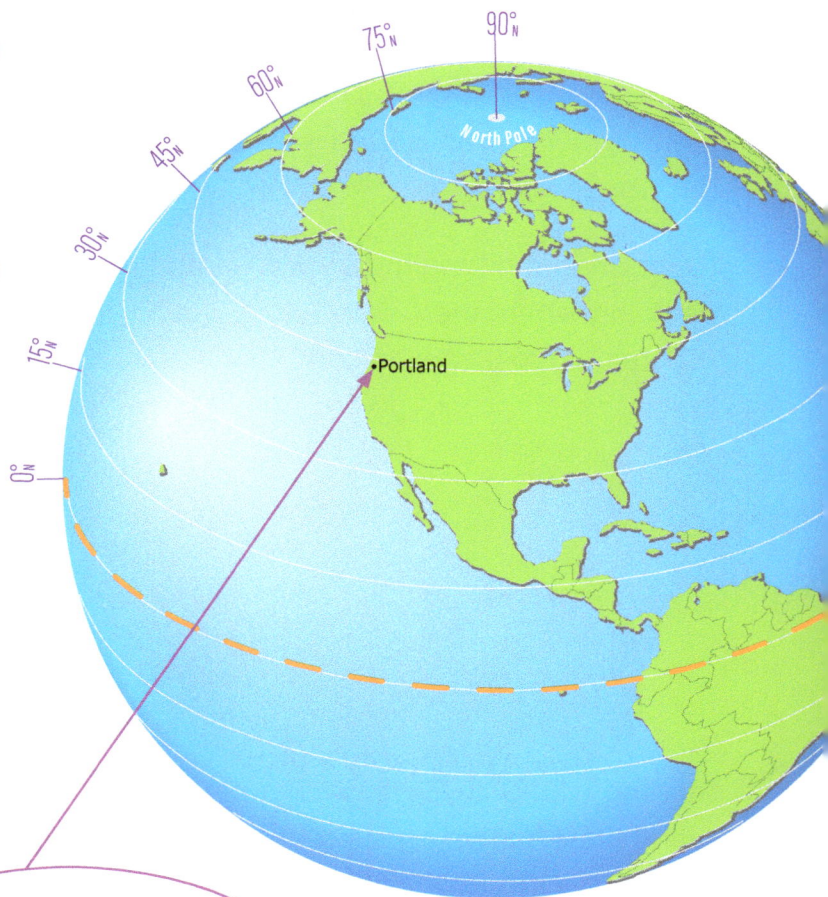

If you lived somewhere around Portland, Oregon, what would your latitude be? Portland is very close to the 45° latitude line and it is north of the equator. So your latitude in Portland would be about 45° N.

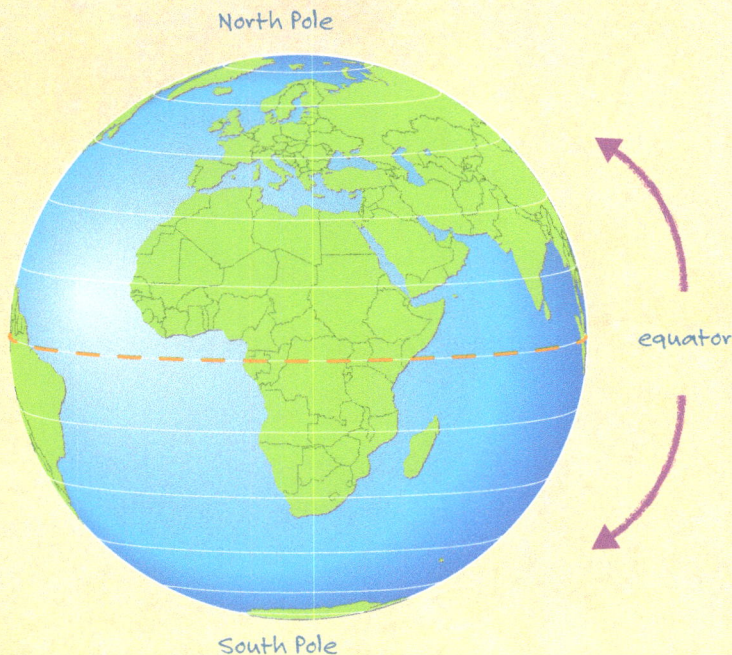

This globe shows the equator, and the latitude lines from there to both the North and South Poles.

41

DEGREES FOR LONGITUDE LINES

Just as latitude lines show how far north or south a place is, longitude lines tell how far east or west it is.

Numbering the longitude lines meant first deciding where the 0° line would be. About 150 years ago it was decided that the 0° longitude line would go through the town of Greenwich (GREN ich) in England.

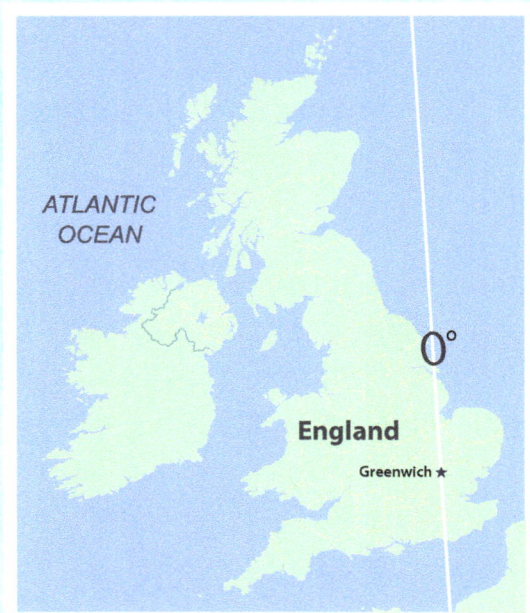

Starting with 0° in Greenwich and going **west** halfway around the earth takes us to the 180° longitude line. These longitude lines, from 0° to 180°, are all marked W for west.

Just like the latitude lines, longitude lines on a globe are usually shown every 15°. Longitude numbers are written along the equator.

To number the longitude lines to the **east** of the 0° line, the numbers start again at 0° and go east to the 180° longitude line. These lines, from 0° to 180°, are all marked E for east.

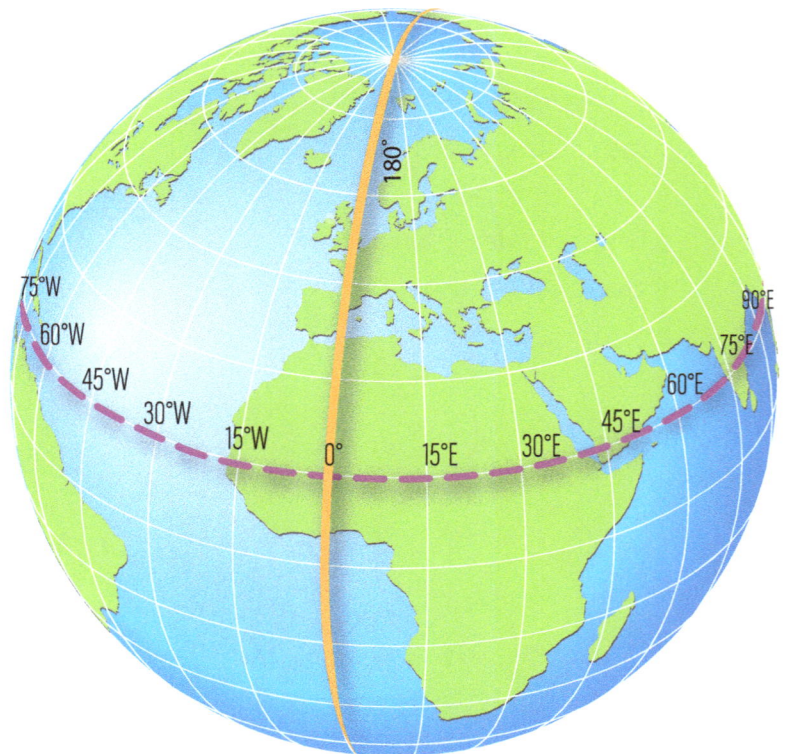

A degree of latitude is always the same distance, no matter where on the globe you are. But this is not true of degrees of longitude.

Lines of longitude are farthest apart at the equator, where the earth is widest. Then they come closer and closer together as they approach the North and South Poles, and gradually meet there.

HEMISPHERES

A **hemisphere** is half a sphere.

To help us describe where things are on the earth, we divide it into two imaginary halves.

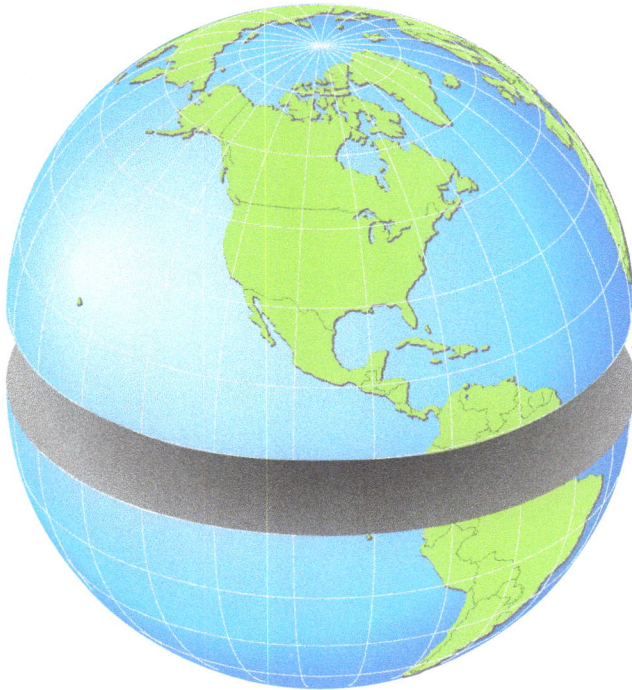

The half from the equator to the North Pole is called the **Northern Hemisphere**.

The half from the equator to the South Pole is called the **Southern** (SUH thurn) **Hemisphere**.

We can also divide the earth into eastern and western halves. The half from 0° to 180° going west is called the **Western Hemisphere**.

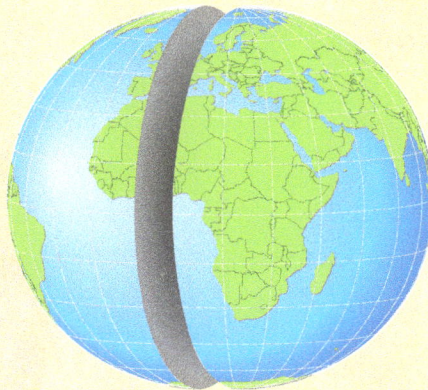

The other half, from 0° to 180° going east is called the **Eastern Hemisphere**.

Let's Do This!

LATITUDE LINES

For this activity you will need

- a globe

- a partner to work with

Steps

1 On your globe, point out the latitude lines.

2 Starting from the equator (latitude 0°) and going north, read aloud the latitude degrees you can find on your globe.

3 Starting from the equator and going south, read aloud the latitude lines you find.

4 Have your partner tell you the degrees of the latitude lines that are marked on the globe north of the equator. For each one, find something on the globe that is at that latitude. Keep going until you can do this easily.

5 Do the same thing with the latitude lines south of the equator.

6 Now have your partner say latitude line degrees north and south of the equator (mixing them up.) Keep going until you can find any latitude line on the globe easily.

LONGITUDE LINES

Let's Do This!

For this activity you will need

- a globe

- a partner to work with

Steps

1 Point out the longitude lines to your partner.

2 Find the 0° longitude line that goes through Greenwich, England.

3 From the 0° longitude line and going west, read out the longitude degrees you can find on your globe until you reach 180°.

4 Start again at the 0° line and go east. Read out the longitude lines until you reach 180°.

5 Have your partner tell you the degrees of the longitude lines west of 0°. For each one, find something on that longitude line. Keep going until you can do this easily.

6 Do the same thing with longitude lines east of 0°.

7 Now have your partner tell you the degrees of longitude lines west and east of 0° (mixing them up.) For each one, find something that is on that longitude line. Keep going until you can find any longitude line on the globe easily.

USING LATITUDE AND

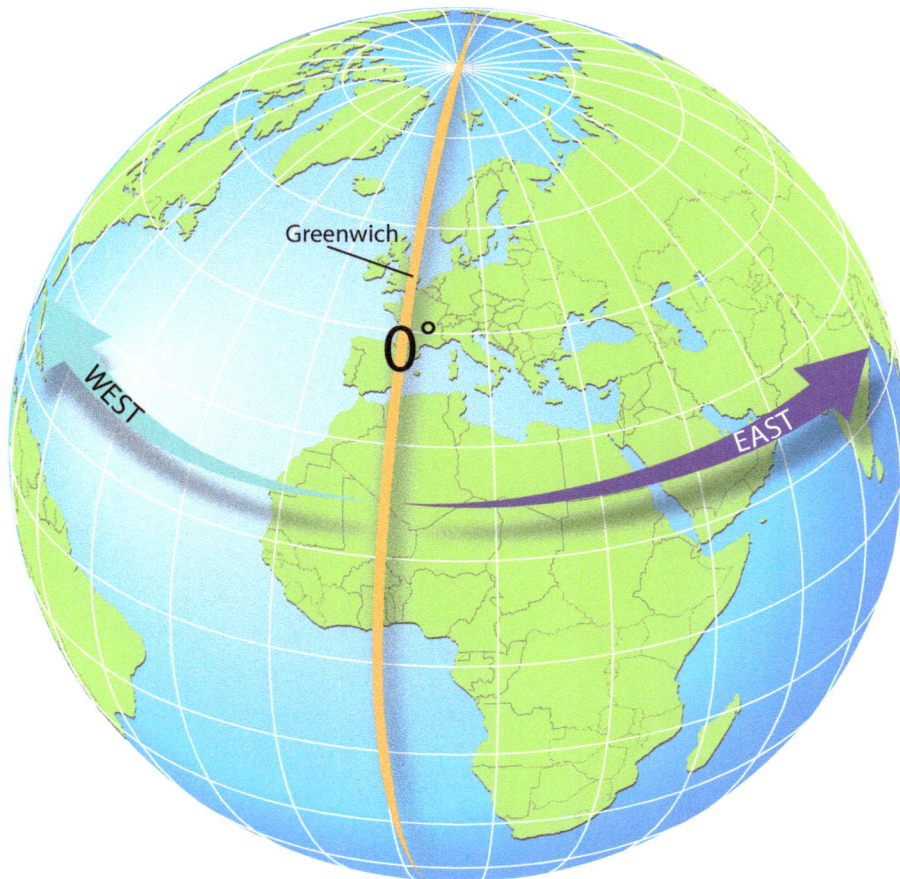

Let's try using the grid of latitude and longitude lines on your globe to find some places on the earth.

Since Greenwich lies right on the 0° longitude line, it is neither east nor west. It's simply 0°. This is true of any place right on the 0° longitude line.

LONGITUDE

Using the grid, let's look at the city of New Orleans in the state of Louisiana.

The latitude of New Orleans is 30° N and its longitude is 90° W. It's located right where the 30° N latitude line and the 90° W longitude lines cross. Written as an "address" this would be 30° N, 90° W.

With only this information, we can easily find New Orleans on a globe. Give it a try.

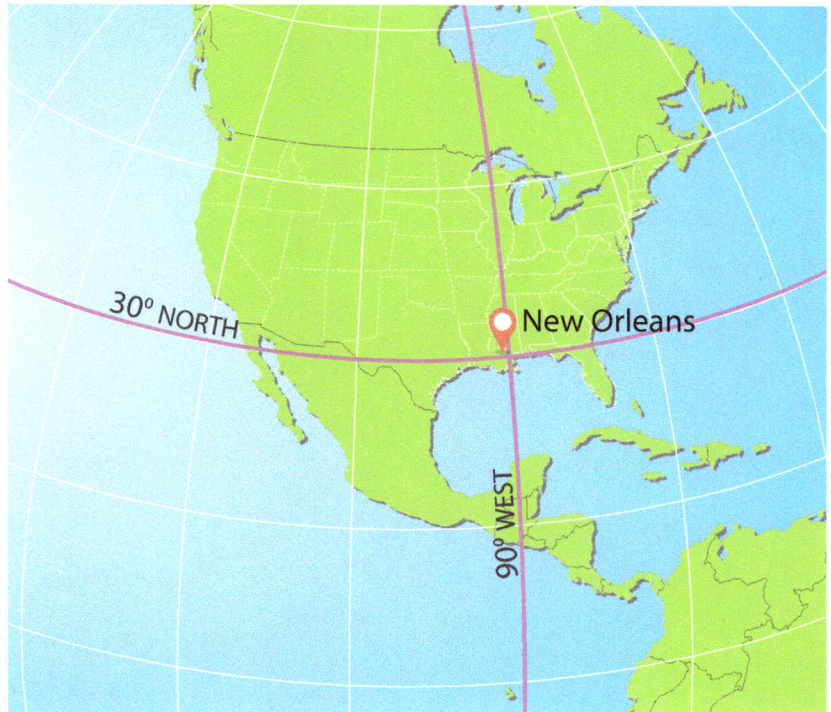

COORDINATES

The latitude and longitude of an "address" are called its **coordinates** (coh ORD nets). Coordinates are a set of numbers used to name an exact place on the globe.

A pair of latitude and longitude coordinates gives us the exact location of any place on Earth. Each number of the address is a single coordinate.

The latitude coordinate is given first and longitude second, with a comma in between. For example, as we saw earlier, New Orleans is 30°N, 90°W.

The latitude coordinate tells you how far north or south a place is. The longitude coordinate tells you how far east or west it is.

For example, use your globe to see if you can discover what South American country, shown by a dot on this globe, is 15° S, 60°W.

Try finding the city with coordinates of 60°N, 30°E. (Hint: it's in Russia!) Notice how far north it is of the equator, and how far east of 0°.

Since globes and maps only show latitude and longitude lines every 15°, you often have to estimate what the coordinates of a place are. They won't be perfect, but they'll be close, and that's fine.

Here's an example. New York City is close to the 75° W longitude line, but not quite on it. You might estimate that its longitude is about 73° W.

It's also just a little north of the 40° N latitude line, so you can estimate that its latitude is about 41° N.

You could say the coordinates of New York City are 41° N, 73° W.

If someone told you they lived at coordinates 42° N, 110° W, you could use latitude and longitude lines to figure out where that is. You might have to guess a bit, but you could get pretty accurate. Can you find it?

As a further example, see if you can find the city with coordinates of 47° N, 122° W.

You can use latitude and longitude coordinates to name or find the location of any place on Earth.

WHERE IS IT?

Let's Do This!

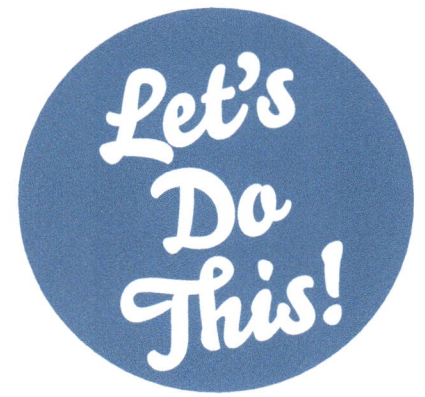

For this activity you will need

- a globe
- a world map
- pencil
- paper
- a partner to work with

Steps

1 On a sheet of paper, make two columns, like this:

Latitude   Longitude

2 In the first column write down all the latitude numbers (N and S) shown on your globe.

3 In the second column write down all the longitude numbers (E and W) shown on your globe.

4 Have your partner tell you one latitude number and one longitude number from your list. You find that location on your globe. Remember, it will be where the latitude and longitude lines cross.

5 Keep going until you can easily find at least five locations in a row.

6 Do the same with a world map. Have your partner tell you one latitude number and one longitude number from your list. Find that location on your map. Remember, it will be where the latitude and longitude lines cross.

7 Keep going until you can easily find at least five locations in a row.

WHAT'S THERE?

For this activity you will need

- a globe

- a world map

- a partner to work with

Steps

1 Find the location of each set of coordinates on your globe. Tell your partner what city is near each location.

- 45° N, 75° W

- 39° N, 9° W

- 30° N, 30° E

- 15° N, 121° E

- 50° N, 0°

- 34° S, 150° E

2 Choose four places on your globe and write down their name and coordinates. Have your partner find each place.

3 Find the location of each set of coordinates on a world map. Tell your partner what city is near each location.

- 49° N, 2° E

- 12° S, 77° W

- 19° N, 99° W

- 61° N, 150° W

- 1° N, 104° E

- 38° S, 145° E

4 Choose four places on the world map and write down their name and coordinates. Have your partner find each place.

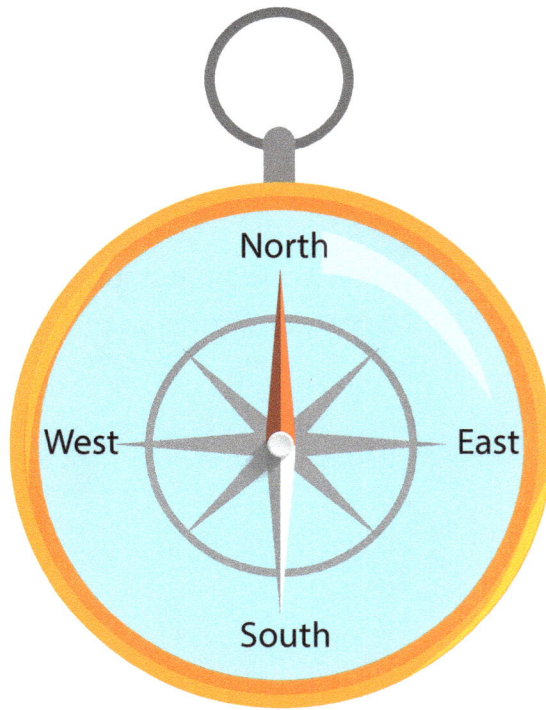

North

West

East

South

MAKE YOUR OWN MAP

For this activity you will need

- a large sheet of paper, roughly 18 inches by 18 inches
- a compass
- colored pencils or crayons
- another person to share your map with

Steps

1. Choose a place you are familiar with. Pick one that is at least the size of a large room and has lots of different features. It can be indoors or outdoors.

2. Work out a map scale that will work for your paper. Make sure to write the scale on your map.

3. Put a compass rose on your map.

4. Draw your map. Before you start, make sure the directions on the compass rose show the actual directions of the place.

5. Put a grid on your map.

6. If you would like, color your map.

7. On a separate piece of paper, make a list of things and places on the map. Put grid numbers and letters next to them. Keep this with the map. (If you want to write them on the back of the map that would be fine.)

8. When your map is complete, show it to another person. Ask them to use the list and grid to find some of the things on the map.

9 CHAPTER GEOGRAPHY

Humans have been learning about the earth for a long time. They know which parts of the world are mountainous, where there are rivers and lakes, how big the oceans are. They've explored it, and learned how to make maps and globes that show us where things are on our big, interesting, amazing planet.

All this is part of a science we call **geography**, which is learning about the earth, what's on it and where those things are.

Now you know a lot more about geography.

And you know a lot more about maps and globes.

You know how to read a map.

You know how to use maps and globes
to find places.

You know how to make simple, useful maps.

This means you've started on the path to
becoming a geographer, someone who knows a
lot about where things are on the earth.

But there are many more fascinating things to find
out about the earth and what's on it.

What can you discover?

You're a young scientist.
Go find out!